Math Advantage

Practice for Standardized Tests

Grade 6

Harcourt Brace & Company

Orlando • Atlanta • Austin • Boston • San Francisco • Chicago • Dallas • New York • Toronto • London

http://www.hbschool.com

▶ Multiple-Choice Formal Practices
Objectives 1–50..1–78

▶ Performance Task Practice
Tasks 1–28...80–135

Copyright © by Harcourt Brace & Company

All rights reserved. No part of this publication may be reproduced or transmitted in any form or by any means, electronic or mechanical, including photocopy, recording, or any information storage and retrieval system.

Teachers using MATH ADVANTAGE may photocopy complete pages in sufficient quantities for classroom use only and not for resale.

HARCOURT BRACE and Quill Design is a registered trademark of Harcourt Brace & Company. MATH ADVANTAGE is a trademark of Harcourt Brace & Company.

Printed in the United States of America

ISBN 0-15-314490-4

2 3 4 5 6 7 8 9 10 073 2000 99 98

Name _____

Practice Objective 1

DIRECTIONS
Read each question and choose the best answer. Then mark the space for the answer you have chosen. If a correct answer is *not here*, mark the space for NH.

SAMPLE
What is 482 rounded to the nearer hundred?

A 400   C 500   E NH
B 480   D 1,000

① The attendance at a football game was 19,853. What is that number rounded to the nearer thousand?

A 10,000   C 19,900   E NH
B 19,000   D 20,000

② The population of Germany is 83,536,115. What is that number rounded to the nearer million?

F 90,000,000
G 86,000,000
H 84,000,000
J 80,000,000
K NH

③ What is the standard form of eighty million, seven hundred five thousand, forty-three?

A 80,705,043
B 80,750,043
C 8,705,043
D 80,705,430
E NH

④ What is the value of 4 in the number 2,947,850?

F Four thousands
G Four ten thousands
H Four millions
J Four hundreds
K NH

⑤ Which number is less than 2,780?

A 2,870   C 3,078   E NH
B 2,980   D 8,002

⑥ Which number is greater than 3,507?

F 3,490   H 2,509   K NH
G 3,315   J 3,705

⑦ Which shows the numbers written from greatest to least?

A 3,665; 3,566; 3,656; 6,356
B 6,356; 3,665; 3,656; 3,566
C 3,566; 3,656; 3,665; 6,356
D 6,356; 3,566; 3,665; 3,656
E NH

⑧ Which shows the numbers written from least to greatest?

F 6,540; 6,504; 5,460; 4,605
G 4,605; 6,504; 6,540; 5,460
H 4,605; 5,460; 6,504; 6,540
J 4,605; 6,504; 5,460; 6,540
K NH

Math Advantage Test Prep

Name _____

Practice Objective 1

DIRECTIONS
Read each question and choose the best answer. Then mark the space for the answer you have chosen. If a correct answer is *not here*, mark the space for NH.

SAMPLE
What is $7.42 rounded to the nearer dollar?

A $7
B $7.40
C $8
D $10
E NH

1 What is $73.59 rounded to the nearer dollar?

A $70
B $73
C $73.60
D $74
E NH

2 Flo Joyner holds the world record for the 100-meter dash. She finished in 10.49 seconds. What is that number rounded to the nearer tenth?

F 10 sec
G 10.4 sec
H 10.5 sec
J 11 sec
K NH

3 The average speed of the winning car in the second Indy 500 race was 78.719 miles per hour. What is that number rounded to the nearer hundredth?

A 79
B 78.71
C 78.72
D 78.7
E NH

4 What is the standard form of fifty thousand, twenty-eight and seventy-one thousandths?

F 50,280.71
G 50,028.71
H 50,028.071
J 50,028.0071
K NH

5 What is the value of 8 in 73.085?

A 8 tens
B 8 hundredths
C 8 thousandths
D 8 tenths
E NH

6 Which number is greater than 5.709?

F 0.795
G 5.09
H 5.7
J 5.099
K NH

7 Which shows the numbers written from greatest to least?

A 5.28, 8.06, 7.53, 8.43
B 8.43, 8.06, 7.53, 5.28
C 8.43, 7.53, 5.28, 8.06
D 5.28, 7.53, 8.06, 8.43
E NH

8 Which shows the numbers written from least to greatest?

F 5.602, 5.026, 5.062, 6.502
G 6.502, 5.602, 5.062, 5.026
H 5.026, 5.062, 5.602, 6.502
J 5.026, 5.062, 6.502, 5.602
K NH

Math Advantage Test Prep

Name _____

Practice Objective 2

DIRECTIONS
Read each question and choose the best answer. Then mark the space for the answer you have chosen. If a correct answer is *not here*, mark the space for NH.

SAMPLE
Which decimal names the same number as 30 + 5 + 0.7 + 0.01?

A 35.71   C 305.71   E NH
B 35.071   D 305.071

① Which decimal names the same number as 40 + 3 + 0.5?

A 43.05   C 43.5   E NH
B 403.05   D 403.5

② Which decimal names the same number as 300 + 70 + 6 + 0.2 + 0.06?

F 376.026   H 376.26   K NH
G 376.206   J 3,076.26

③ Which decimal names the same number as 900 + 5 + 0.7 + 0.01?

A 950.071   C 905.701   E NH
B 905.71   D 905.0701

④ Which decimal names the same number as 2,000 + 60 + 3 + 0.8?

F 2,063.8   H 2,063.008   K NH
G 2,063.08   J 2,603.008

⑤ Which gives 324.809 in expanded form?

A 300 + 20 + 4 + 0.08 + 0.9
B 300 + 20 + 4 + 0.8 + 0.09
C 300 + 20 + 4 + 0.08 + 0.009
D 300 + 20 + 4 + 0.8 + 0.009
E NH

⑥ Which gives 407.092 in expanded form?

F 400 + 7 + 0.9 + 0.002
G 400 + 70 + 0.09 + 0.002
H 400 + 7 + 0.09 + 0.002
J 40 + 7 + 0.9 + 0.02
K NH

⑦ Which gives 23,050.407 in expanded form?

A 20,000 + 3,000 + 50 + 0.04 + 0.7
B 20,000 + 3,000 + 50 + 0.4 + 0.07
C 20,000 + 3,000 + 50 + 0.4 + 0.007
D 20,000 + 3,000 + 50 + 0.04 + 0.007
E NH

⑧ Which gives 45,006.019 in expanded form?

F 40,000 + 500 + 6 + 0.01 + 0.009
G 40,000 + 5,000 + 600 + 0.1 + 0.09
H 40,000 + 5,000 + 6 + 0.01 + 0.009
J 4,000 + 500 + 6 + 0.1 + 0.09
K NH

Math Advantage Test Prep

Name _____

DIRECTIONS

Read each question and choose the best answer. Then mark the space for the answer you have chosen. If a correct answer is *not here,* mark the space for NH.

SAMPLE

Carol made these bank deposits in December: $89, $191, and $92. *About* how much did she deposit in December?

A About $480
B About $400
C About $300
D About $100

1 Carmen traveled 3,949 miles on vacation. Geraldine traveled 1,258 miles. *About* how much farther did Carmen travel than Geraldine?

A About 1,000 miles
B About 3,000 miles
C About 4,000 miles
D About 5,000 miles

2 There are 32 watercoolers in an office building. Each watercooler holds 26 gallons of water. *About* how many gallons are there in all?

F About 60 gallons
G About 900 gallons
H About 6,000 gallons
J About 9,000 gallons

3 The librarians put 3,281 books on shelves at the library. They put 27 books on each shelf. *About* how many shelves did they need?

A About 90 shelves
B About 100 shelves
C About 150 shelves
D About 200 shelves

4 Sasha read one book with 325 pages, another with 549 pages, and another with 908 pages. *About* how many pages did she read?

F About 3,000 pages
G About 2,500 pages
H About 1,700 pages
J About 1,300 pages

5 There are 5,356 books in the school library. Of these books, 4,219 can be loaned to students. *About* how many books cannot be loaned to students?

A About 2,000 books
B About 1,000 books
C About 800 books
D About 500 books

6 A glacier can move about 225 inches in 74 years. *About* how far can it move in one year?

F About 3 inches
G About 30 inches
H About 300 inches
J About 3,000 inches

4

Math Advantage Test Prep

Name _____

Practice Objective 3

DIRECTIONS
Read each question and choose the best answer. Then mark the space for the answer you have chosen. If a correct answer is *not here,* mark the space for NH.

SAMPLE

A large department store has 65 western hats, 42 straw hats, 71 baseball caps, and 36 dressy hats for sale. *About* how many hats did they have?

A About 150 hats
B About 200 hats
C About 250 hats
D About 300 hats

1 Neka has 128 blue marbles, 36 green marbles, 72 red marbles, and 63 yellow marbles. *About* how many marbles does she have in all?

A About 200 marbles
B About 250 marbles
C About 300 marbles
D About 150 marbles

2 In June, 250 people went to the park. In July, 850 people went to the park, and 750 people went to the park in August. *About* how many people went to the park during these months?

F About 1,600 people
G About 1,850 people
H About 2,000 people
J About 2,850 people

3 The Rodriguez family traveled 3,538 miles in 57 hours. *About* how many miles did they travel each hour?

A About 240,000 miles
B About 6,000 miles
C About 600 miles
D About 60 miles

4 Twenty-two buses were needed to take 1,276 fans to the basketball playoffs. *About* how many people rode in each bus?

F About 50 people
G About 60 people
H About 500 people
J About 600 people

5 Darnell scored these points on four quizzes. *About* how many points did he score in all?
41   36   42   44

A About 120 points
B About 140 points
C About 160 points
D About 200 points

6 Keisha ran these distances in three days. *About* how far did she run altogether?
815 yards   790 yards   792 yards

F About 1,800 yards
G About 2,100 yards
H About 2,400 yards
J About 3,600 yards

Math Advantage Test Prep

Name _____

Practice Objective 4

DIRECTIONS
Read each question and choose the best answer. Then mark the space for the answer you have chosen. If a correct answer is *not here,* mark the space for NH.

SAMPLE

Jimmy bought 3.9 pounds of peanuts, 1.9 pounds of potato chips, and 4.8 pounds of pretzels. How many pounds of snack food did he buy?

Estimate. Then decide which answer is most reasonable.

- A 12.8 pounds
- B 10.6 pounds
- C 9.3 pounds
- D 8.7 pounds

1 During each daily gym class, Jeannie runs around the track 11 times. How many times will she run around the track in a school year with 172 days?

Estimate. Then decide which answer is most reasonable.

- A 1,458 times
- B 1,892 times
- C 2,845 times
- D 3,025 times

2 Jill clipped 3,356 recipes in 1998 and 2,748 recipes in 1999. Karen clipped a total of 4,219 recipes. How many more recipes did Jill clip than Karen?

Estimate. Then decide which answer is most reasonable.

- F 2,942 recipes
- G 1,885 recipes
- H 10,221 recipes
- J 1,043 recipes

3 Patti hiked $6\frac{3}{4}$ miles on Saturday and $2\frac{7}{8}$ miles on Sunday. How much farther did she hike on Saturday than on Sunday?

Estimate. Then decide which answer is most reasonable.

- A $9\frac{5}{8}$ miles
- B $5\frac{1}{8}$ miles
- C $8\frac{3}{8}$ miles
- D $3\frac{7}{8}$ miles

4 At the apple orchard, John picked $13\frac{7}{8}$ pounds of red delicious apples and $9\frac{1}{2}$ pounds of gala apples. How many pounds of apples did he pick all together?

Estimate. Then decide which answer is most reasonable.

- F $4\frac{3}{8}$ lb
- G $4\frac{1}{4}$ lb
- H $22\frac{3}{4}$ lb
- J $23\frac{3}{8}$ lb

5 Lizzie's bowling score for one game is usually between 80 and 110. What is a reasonable answer for her point total in 8 games?

- A About 550 points
- B About 750 points
- C About 900 points
- D About 1,000 points

6 Flannel fabric is sale priced at $7.15 for 2 yards. What is a reasonable estimate for the cost of 19 yards of flannel fabric?

- F About $100
- G About $70
- H About $60
- J About $45

Math Advantage Test Prep

Name _____

Practice Objective 5

DIRECTIONS
Read each question and choose the best answer. Then mark the space for the answer you have chosen. If a correct answer is *not here,* mark the space for NH.

SAMPLE

4.5
$+3.29$

A 7.29   C 7.709   E NH
B 7.7   D 7.79

1 3.29
    $+7.48$

A 10.77   C 11.77   E NH
B 10.67   D 1,177

2 $61.87 + 3.2 + 7.09$

F 7,216   H 69.28   K NH
G 71.16   J 16.477

3 20.31
    -18.64

A 1.67   C 12.57   E NH
B 10.67   D 38.95

4 $57.3 - 38.65$

F 18.65   H 32.92   K NH
G 29.75   J 95.95

5 $6 - 1.38$

A 7.38   C 4.62   E NH
B 5.38   D 4.38

6 6.2
    $\times 2.5$

F 1,550   H 15.5   K NH
G 155   J 15.4

7 0.71×3.8

A 0.2698   C 26.98   E NH
B 2.698   D 269.8

8 2.05
    $\times 1.66$

F 34,030   H 34.03   K NH
G 340.3   J 0.3403

9 Sally buys 3.4 pounds of ground beef, 2.8 pounds of ground pork, and 1.7 pounds of ground veal. How many pounds of ground meat does she buy?

A 6 pounds
B 6.9 pounds
C 7.9 pounds
D 8 pounds
E NH

Math Advantage Test Prep

Name _____

Practice Objective 6

DIRECTIONS
Read each question and choose the best answer. Then mark the space for the answer you have chosen. If a correct answer is *not here*, mark the space for NH.

SAMPLE
$4\overline{)26.8}$

A 0.67   C 67   E NH
B 6.7   D 6.2

① $8\overline{)108.8}$

A 1.36   C 136   E NH
B 13.6   D 11.1

② $6\overline{)16.02}$

F 3.67   H 2.6   K NH
G 20.67 J 267

③ $18\overline{)52.2}$

A 290   C 3.2   E NH
B 29   D 2.9

④ $0.6\overline{)2.28}$

F 38   H 3.8   K NH
G 0.38 J 3.08

⑤ $2.3\overline{)9.384}$

A 4.08 C 4.8   E NH
B 480   D 408

⑥ $1.4\overline{)2.086}$

F 14.9 H 1.49 K NH
G 149   J 0.149

⑦ $3.3\overline{)27.192}$

A 0.824 C 824   E NH
B 82.4   D 8.24

⑧ $4.9\overline{)22.589}$

F 0.0461 H 0.0461 K NH
G 461   J 4.61

⑨ Jim and three friends equally share $3.68. How much money does each person get?

A $ 0.92
B $14.72
C $9.20
D $11.04
E NH

⑩ Cindy has a piece of lumber that is 14.4 meters long. She wants to make shelves for her room. Each shelf will be 2.4 meters long. How many shelves can she cut from the piece of lumber?

F 5 shelves
G 6 shelves
H 7 shelves
J 35 shelves
K NH

Math Advantage Test Prep

Name _____

Practice Objective 7

DIRECTIONS
Read each question and choose the best answer. Then mark the space for the answer you have chosen. If a correct answer is *not here,* mark the space for NH.

SAMPLE
Which fraction is greater than $\frac{1}{3}$?

A $\frac{1}{8}$ C $\frac{2}{6}$ E NH
B $\frac{2}{9}$ D $\frac{2}{3}$

① Which fraction is greater than $\frac{4}{9}$?

A $\frac{1}{4}$ C $\frac{8}{18}$ E NH
B $\frac{1}{18}$ D $\frac{4}{10}$

② Which fraction is less than $\frac{2}{3}$?

F $\frac{10}{15}$ H $\frac{7}{10}$ K NH
G $\frac{3}{4}$ J $\frac{3}{5}$

③ Which mixed number is greater than $2\frac{3}{4}$?

A $2\frac{4}{9}$ C $2\frac{9}{10}$ E NH
B $2\frac{3}{5}$ D $1\frac{4}{5}$

④ Which mixed number is less than $20\frac{1}{10}$?

F $20\frac{9}{10}$ H $20\frac{1}{20}$ K NH
G $30\frac{1}{100}$ J $20\frac{1}{5}$

⑤ Which shows the fractions written from least to greatest?

A $\frac{1}{2}, \frac{1}{3}, \frac{1}{4}, \frac{1}{10}$ D $\frac{1}{2}, \frac{1}{3}, \frac{1}{4}, \frac{1}{10}$
B $\frac{1}{4}, \frac{1}{3}, \frac{1}{2}, \frac{1}{10}$ E NH
C $\frac{1}{10}, \frac{1}{4}, \frac{1}{3}, \frac{1}{2}$

⑥ Which shows the fractions written from greatest to least?

F $\frac{9}{10}, \frac{3}{5}, \frac{3}{10}, \frac{1}{2}$ J $\frac{1}{2}, \frac{3}{5}, \frac{3}{10}, \frac{9}{10}$
G $\frac{9}{10}, \frac{3}{5}, \frac{1}{2}, \frac{3}{10}$ K NH
H $\frac{3}{10}, \frac{1}{2}, \frac{3}{5}, \frac{9}{10}$

⑦ Which is the greatest mixed number?
$2\frac{5}{8}, 2\frac{3}{4}, 2\frac{1}{2}, 2\frac{7}{16}, 2\frac{1}{4}$

A $2\frac{5}{8}$ C $2\frac{1}{2}$ E $2\frac{1}{4}$
B $2\frac{3}{4}$ D $2\frac{7}{16}$

⑧ Which shows the mixed numbers written from greatest to least?

F $12\frac{3}{4}, 13\frac{1}{2}, 12\frac{1}{3}, 13\frac{2}{5}$
G $13\frac{2}{5}, 13\frac{1}{2}, 12\frac{3}{4}, 12\frac{1}{3}$
H $13\frac{1}{2}, 13\frac{2}{5}, 12\frac{3}{4}, 12\frac{1}{3}$
J $12\frac{1}{3}, 12\frac{3}{4}, 13\frac{2}{5}, 13\frac{1}{2}$
K NH

Math Advantage Test Prep

Name _____

Practice Objective 7

DIRECTIONS
Read each question and choose the best answer. Then mark the space for the answer you have chosen. If a correct answer is *not here,* mark the space for NH.

SAMPLE
Which fraction is equivalent to $\frac{1}{4}$?

A $\frac{2}{10}$ C $\frac{2}{8}$ E NH
B $\frac{4}{12}$ D $\frac{8}{16}$

1 Which fraction is equivalent to $\frac{3}{8}$?

A $\frac{6}{10}$ C $\frac{6}{16}$ E NH
B $\frac{8}{16}$ D $\frac{6}{24}$

2 What is the missing number?
$4\frac{7}{10} = \frac{n}{10}$

F 47   H 38   K NH
G 41   J 21

3 What is the missing number?
$8\frac{3}{4} = \frac{n}{4}$

A 15   C 27   E NH
B 20   D 29

4 Which fraction is equivalent to $\frac{12}{20}$?

F $\frac{12}{40}$ H $\frac{24}{20}$ K NH
G $\frac{24}{30}$ J $\frac{3}{5}$

5 Emmy has finished $\frac{9}{12}$ of a rug she is making. Which fraction is the simplest form of $\frac{9}{12}$?

A $\frac{2}{3}$ C $\frac{18}{24}$ E NH
B $\frac{3}{4}$ D $\frac{12}{16}$

6 The model shows how much pizza Kirk had left for his party after he ate a slice. Which fraction could you use to show this?

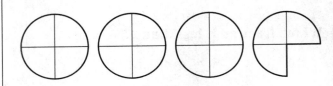

F $4\frac{1}{4}$ H $3\frac{1}{4}$ K NH
G $\frac{4}{15}$ J $\frac{15}{8}$

7 Mary has $3\frac{1}{4}$ loaves of bread. Which fraction is another name for $3\frac{1}{4}$?

A $\frac{13}{3}$ C $\frac{7}{4}$ E NH
B $\frac{13}{4}$ D $\frac{4}{13}$

8 Kyle has $\frac{25}{8}$ pizza. What is $\frac{25}{8}$ written as a mixed number?

F $2\frac{1}{8}$ H $3\frac{1}{8}$ K NH
G $4\frac{1}{8}$ J $8\frac{1}{3}$

Math Advantage Test Prep

Name _____

Practice Objective 8

DIRECTIONS

Read each question and choose the best answer. Then mark the space for the answer you have chosen. If a correct answer is *not here*, mark the space for NH.

SAMPLE

What is the GCF (greatest common factor) of 12 and 16?

A 2   C 6   E NH
B 4   D 8

① What is the GCF of 9 and 12?

A 3   C 9   E NH
B 4   D 12

② What is the GCF of 28 and 16?

F 2   H 8   K NH
G 4   J 16

③ What is the GCF of 32 and 24?

A 2   C 8   E NH
B 4   D 16

④ What is the GCF of 21 and 35?

F 3   H 7   K NH
G 5   J 21

⑤ Choose the simplest form for $\frac{12}{20}$.

A $\frac{3}{10}$ C $\frac{2}{3}$ E NH
B $\frac{6}{10}$ D $3\frac{1}{3}$

⑥ Choose the simplest form for $\frac{15}{35}$.

F $\frac{3}{8}$ H $\frac{3}{7}$ K NH
G $\frac{1}{3}$ J $\frac{5}{7}$

⑦ Choose the simplest form for $\frac{48}{64}$.

A $\frac{2}{3}$ C $\frac{6}{8}$ E NH
B $\frac{3}{4}$ D $\frac{12}{16}$

⑧ Choose the simplest form for $\frac{24}{32}$.

F $\frac{3}{5}$ H $\frac{3}{4}$ K NH
G $\frac{2}{3}$ J $\frac{5}{6}$

⑨ Choose the simplest form for $\frac{35}{40}$.

A $\frac{7}{8}$ C $\frac{5}{7}$ E NH
B $\frac{4}{5}$ D $\frac{7}{9}$

⑩ Choose the simplest form for $\frac{36}{48}$.

F $\frac{2}{3}$ H $\frac{12}{16}$ K NH
G $\frac{3}{4}$ J $\frac{18}{24}$

⑪ Choose the simplest form for $\frac{32}{56}$.

A $\frac{1}{2}$ C $\frac{16}{28}$ E NH
B $\frac{4}{7}$ D $\frac{8}{14}$

Math Advantage Test Prep

Name _____

Practice Objective 8

DIRECTIONS
Read each question and choose the best answer. Then mark the space for the answer you have chosen. If a correct answer is *not here*, mark the space for NH.

SAMPLE
What is the LCM (least common multiple) of 4 and 6?

A 4   C 12   E NH
B 6   D 24

① What is the LCM of 2 and 3?

A 3   C 12   E NH
B 6   D 24

② What is the LCM of 8 and 6?

F 16   H 24   K NH
G 18   J 36

③ Which pair of numbers has 24 as the LCM?

A 2, 12   C 3, 6   E NH
B 3, 8   D 24, 28

④ The denominators of $\frac{3}{8}$ and $\frac{1}{3}$ are 8 and 3. What is their LCM?

F 12   H 24   K NH
G 18   J 48

⑤ The denominators of $\frac{3}{8}$ and $\frac{5}{16}$ are 8 and 16. What is their LCM?

A 16   C 4   E NH
B 8   D 2

⑥ The denominators of $\frac{2}{3}$ and $\frac{3}{10}$ are 3 and 10. What is their LCM?

F 3   H 30   K NH
G 10   J 60

⑦ The denominators of $\frac{11}{15}$ and $\frac{2}{5}$ are 15 and 5. What is their LCM?

A 5   C 20   E NH
B 10   D 25

⑧ Find $1\frac{7}{8} - \frac{3}{4}$.

F $\frac{1}{8}$ H 1   K NH
G $2\frac{5}{8}$ J $1\frac{1}{8}$

⑨ Find $2\frac{3}{16} - \frac{2}{3}$.

A $1\frac{25}{48}$ C $2\frac{23}{48}$ E NH
B $1\frac{1}{2}$ D $2\frac{41}{48}$

⑩ Find $2\frac{11}{12} + 1\frac{7}{8}$.

F $1\frac{19}{24}$ H $1\frac{1}{24}$ K NH
G $4\frac{19}{24}$ J $3\frac{19}{24}$

12   Math Advantage Test Prep

Name _____

Practice Objective 8

DIRECTIONS
Read each question and choose the best answer. Then mark the space for the answer you have chosen. If a correct answer is *not here,* mark the space for NH.

SAMPLE

$$\frac{1}{2}$$
$$+\frac{2}{3}$$

A $\frac{2}{5}$ C $\frac{1}{6}$ E NH
B $\frac{5}{6}$ D $\frac{2}{3}$

1
$$\frac{1}{8}$$
$$+\frac{1}{4}$$

A $\frac{3}{8}$ C $\frac{3}{12}$ E NH
B $\frac{1}{6}$ D $\frac{3}{16}$

2
$$9\frac{7}{8}$$
$$+1\frac{2}{3}$$

F $10\frac{13}{24}$ H $11\frac{1}{8}$ K NH
G $1\frac{13}{24}$ J $11\frac{13}{24}$

3
$$\frac{5}{8}$$
$$-\frac{1}{4}$$

A $\frac{3}{8}$ C $\frac{1}{2}$ E NH
B $\frac{7}{8}$ D $\frac{1}{8}$

4
$$8\frac{7}{16}$$
$$-3\frac{1}{4}$$

F $11\frac{11}{16}$ H $5\frac{3}{8}$ K NH
G $5\frac{11}{16}$ J $4\frac{3}{16}$

5
$$15\frac{11}{12}$$
$$-3\frac{5}{8}$$

A $19\frac{13}{24}$ C $13\frac{13}{24}$ E NH
B $12\frac{7}{24}$ D $12\frac{1}{3}$

6 Carrie buys $3\frac{1}{2}$ lb of red potatoes and $4\frac{1}{8}$ lb of Idaho potatoes. How many pounds of potatoes does she buy in all?

F $7\frac{5}{8}$ lb   H 8 lb   K NH
G $7\frac{1}{4}$ lb   J $\frac{5}{8}$ lb

7 Darl runs $1\frac{3}{5}$ mile. Cassie runs $2\frac{1}{5}$ miles. How much farther does Cassie run than Darl?

A $3\frac{4}{5}$ mi   C $\frac{3}{5}$ mi   E NH
B $1\frac{2}{5}$ mi   D 1 mi

Math Advantage Test Prep

Name _____

Practice Objective 9

DIRECTIONS
Read each question and choose the best answer. Then mark the space for the answer you have chosen. If a correct answer is *not here*, mark the space for NH.

SAMPLE
$\frac{1}{2} \times \frac{5}{8}$

A $\frac{5}{16}$ C $\frac{5}{10}$ E NH
B $\frac{10}{16}$ D $\frac{6}{10}$

1 $\frac{2}{5} \times \frac{3}{8}$

A $\frac{6}{20}$ C $6\frac{2}{3}$ E NH
B $\frac{5}{40}$ D $\frac{3}{40}$

2 $6 \times 4\frac{1}{3}$

F 26   H $25\frac{1}{3}$ K NH
G 25   J 24

3 $2\frac{1}{2} \times 3\frac{2}{5}$

A $9\frac{1}{2}$ C $8\frac{1}{2}$ E NH
B 9   D $6\frac{1}{5}$

4 $\frac{1}{2} \div 6$

F $\frac{1}{12}$ H 3   K NH
G $\frac{1}{3}$ J 12

5 $\frac{2}{3} \div \frac{8}{15}$

A $\frac{4}{5}$ C $1\frac{1}{5}$ E NH
B $\frac{16}{45}$ D $1\frac{1}{4}$

6 $2\frac{1}{3} \div \frac{1}{2}$

F $\frac{3}{14}$ H $1\frac{1}{6}$ K NH
G $\frac{6}{7}$ J $5\frac{1}{3}$

7 Mel had 32 sets of baseball cards. He sold $\frac{3}{8}$ of them. How many sets of cards did he sell?

A 4 sets   C 12 sets   E NH
B 9 sets   D 16 sets

8 Carla's recipe calls for $4\frac{1}{2}$ teaspoons of sugar. Since she is making only half the recipe, she only needs $\frac{1}{2}$ of that amount. How much sugar does Carla need to use?

F $2\frac{1}{8}$ teaspoons
G $2\frac{1}{4}$ teaspoons
H 5 teaspoons
J 9 teaspoons
K NH

9 Denise has a $10\frac{1}{2}$-inch board. She cuts it into $\frac{3}{4}$-inch pieces. How many pieces does she get?

A $7\frac{3}{4}$ pieces
B 8 pieces
C 14 pieces
D $14\frac{1}{4}$ pieces
E NH

Math Advantage Test Prep

Name _____

Practice Objective 10

DIRECTIONS

Read each question and choose the best answer. Then mark the space for the answer you have chosen. If a correct answer is *not here,* mark the space for NH.

SAMPLE

Which number is a composite number?
5, 13, 24, 29

A 5   C 24   E NH
B 13 D 29

1 Which number is a prime number?
13, 40, 49, 69

A 13   C 49   E NH
B 40   D 69

2 Which number is a composite number?
17, 27, 53, 31

F 17   H 31   K NH
G 27   J 53

3 Which number is a prime number?
15, 21, 39, 55

A 15   C 39   E NH
B 21   D 55

4 Which number is a composite number?
3, 5, 17, 21

F 3   H 17   K NH
G 5   J 21

5 Which list shows all the prime numbers between 30 and 40?

A 31, 37
B 31, 33, 37, 39
C 31, 35
D 33, 37
E 31, 33, 35, 37, 39

6 Which list shows all the prime numbers less than 10?

F 1, 3, 5, 7, 9
G 3, 5, 7, 9
H 7
J 2, 3, 5, 7
K 3, 5, 7

7 Which list shows all the prime numbers between 50 and 60?

A 51, 53, 55, 57, 59
B 51, 53, 59
C 53, 59
D 53, 57, 59
E 53, 57

8 Which list shows all the prime numbers between 70 and 80?

F 71, 73, 75, 77, 79
G 71, 73, 77
H 71, 73, 79
J 73, 79
K 75, 77

Math Advantage Test Prep

Name _____

Practice Objective 11

DIRECTIONS
Read each question and choose the best answer. Then mark the space for the answer you have chosen. If a correct answer is *not here,* mark the space for NH.

SAMPLE
Which is the greatest integer?

A $^-2$ C 3 E $^-9$
B 0 D $^-4$

1 Which is less than 0?

A $^-3$ C 0 E 1
B 2 D 3

2 Which is greater than 0?

F $^-10$ H $^-9$ K 0
G $^+4$ J $^-3$

3 What is the opposite of 5?

A 5 C $|5|$ E NH
B $^-5$ D $\frac{1}{5}$

4 Which of these shows the integers written in order from least to greatest?

F $^-1, ^-2, 0, 4$
G $4, ^-2, ^-1, 0$
H $^-2, ^-1, 0, 4$
J $0, ^-1, ^-2, 4$
K NH

5 Which of these shows the integers written in order from greatest to least?

A $^-8, ^-6, 4, ^-2$
B $^-8, ^-6, ^-2, 4$
C $4, ^-2, ^-6, ^-8$
D $^-8, 4, ^-2, ^-6$
E NH

6 Marisa's scores for five rounds were $^-9, ^-8, 7, 6,$ and 1. Order the numbers from greatest to least.

F $^-8, 6, 7, 1, ^-9$
G $^-8, 1, 6, 7, ^-9$
H $^-9, ^-8, 1, 6, 7$
J $7, 6, 1, ^-9, ^-8$
K NH

7 Ian recorded these low temperatures in degrees Fahrenheit for 5 days.
$4°F, ^-3°F, 2°F, ^-1°F, 0°F$
Order the numbers from least to greatest.

A $4, ^-3, 2, ^-1, 0$
B $4, 2, 0, ^-1, ^-3$
C $^-3, ^-1, 0, 2, 4$
D $0, ^-1, 2, ^-3, 4$
E $0, ^-1, ^-3, 2, 4$

Math Advantage Test Prep

Name _____

Practice Objective 12

DIRECTIONS

Read each question and choose the best answer. Then mark the space for the answer you have chosen. If a correct answer is *not here,* mark the space for NH.

SAMPLE
$^-13 + {}^+7 =$

A $^-20$ C 6 E NH
B $^-6$ D 20

5 $^+7 - {}^+12 =$

A $^+19$ C $^-5$ E NH
B $^+5$ D $^-19$

1 $^-10 + {}^+5 =$

A $^-15$ C $^+10$ E NH
B $^+5$ D $^+15$

6 $^-4 - {}^+11 =$

F $^+15$ H $^-7$ K NH
G $^+7$ J $^-15$

2 $^+8 + {}^-6 =$

F $^-14$ H $^+2$ K NH
G $^-2$ J $^+14$

7 $^+14 - {}^-11 =$

A $^+23$ C $^-3$ E NH
B $^+3$ D $^-23$

3 $^-24 + {}^-9 =$

A $^-33$ C $^+15$ E NH
B $^-15$ D $^+33$

8 During the football game, the home team lost 4 yards on the first play. On the next play they gained 15 yards. Describe the total change of the home team's position using integers.

F $^-4 + {}^-15 = {}^-19$
G $^-4 + 15 = {}^+11$
H $^+4 + {}^-15 = {}^-11$
J $^+4 + 15 = {}^+19$
K $^-4 - {}^+15 = {}^-19$

4 The temperature was $^-5°C$ at noon. By sunset, the temperature had dropped 12°. What was the temperature at sunset?

F $^-17°C$ H $7°C$ K NH
G $^-7°C$ J $17°C$

Math Advantage Test Prep

17

Name _____

Practice Objective 13

DIRECTIONS

Read each question and choose the best answer. Then mark the space for the answer you have chosen. If a correct answer is *not here*, mark the space for NH.

SAMPLE

What is $\frac{3}{4}$ written as a percent?

A 34%   C 60%   E NH
B 43%   D 75%

Use the figure for questions 1–3.

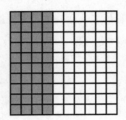

1. What decimal shows the part of the whole square that is shaded?

   A 0.040   C 0.6   E NH
   B 0.40   D 40

2. What fraction of the figure is shaded? Write the fraction in lowest terms.

   F $\frac{40}{60}$ H $\frac{2}{5}$ K NH
   G $\frac{1}{2}$ J $\frac{4}{100}$

3. What percent of the squares are shaded?

   A 4%   C 44%   E NH
   B 40%   D 140%

Use the figure for questions 4–6.

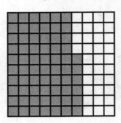

4. What decimal shows the part of the square that is shaded?

   F 6.6   H 0.34   K NH
   G 0.65   J 0.066

5. What fraction of the figure is shaded? Write the fraction in lowest terms.

   A $\frac{33}{100}$ C $\frac{33}{50}$ E NH
   B $\frac{34}{100}$ D $\frac{2}{3}$

6. What percent of the figure is shaded?

   F 660%   H 34%   K NH
   G 66%   J 6.6%

7. What percent of the tile floor is gray?

   A 15%   C 30%   E NH
   B 20%   D $33\frac{1}{3}$%

18

Math Advantage Test Prep

Name _____

Practice Objective 14

DIRECTIONS
Read each question and choose the best answer. Then mark the space for the answer you have chosen.

SAMPLE
Suppose you bought items for $2.75, $1.50, and $8.75. You gave the clerk a $20 bill. Which calculator key sequence would give you the correct change?

A 20 [−] 2.75 [+] 1.50 [+] 8.75 [=]
B 2.75 [+] 1.50 [+] 8.75 [−] 20 [=]
C 20 [−] 2.75 [−] 1.50 [−] 8.75 [=]
D 20 [−] 2.75 [+] 1.50 [+] 8.75 [=]

2 Mel bought a $50 jacket on sale. The discount was 30%. To find the amount of the discount, which key sequence would you use?

F 50 [×] 0.30 [=]
G 30 [×] 0.05 [=]
H 0.30 [×] 50 [×] 15 [=]
J 0.50 [×] 30 [×] 15 [=]

1 A machine fills 8 bottles every 10 seconds. To find how many bottles can be filled in 35 seconds, Jane wrote this proportion: $\frac{8}{10} = \frac{n}{35}$. Which key sequence should she use to solve the proportion?

A 8 [+] 35 [÷] 10 [=]
B 8 [×] 35 [×] 10 [=]
C 8 [×] 35 [−] 10 [=]
D 8 [×] 35 [÷] 10 [=]

3 Lorenzo used a calculator to change $\frac{5}{8}$ to a decimal. What did he do?

A Divide the denominator by the numerator.
B Multiply the numerator by the denominator.
C Divide the numerator by the denominator.
D Multiply the denominator by the numerator.

4 On a simple four-function calculator, Dennis used the key strokes

3 [×] 3 [=] [=] [=] [=]

to find which of the following?

F 9^4
G 3^5
H 9×4
J 3^4

Math Advantage Test Prep

19

Name _____

Practice Objective 15

DIRECTIONS
Read each question and choose the best answer. Then mark the space for the answer you have chosen.

SAMPLE
Which Venn diagram shows the relationship between the vowels {a, e, i, o, u} and the letters of the alphabet?

A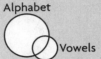

B

C

D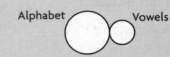

1 What does the Venn diagram show?

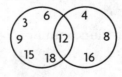

A 12 is a multiple of 4 or a multiple of 3.
B 12 is a common multiple of 3 and 4.
C 16 is a multiple of 3.
D 18 is a multiple of 4.

2 Choose the word that best completes this statement.

{2, 4, 6} is ____?____ of {1, 2, 3, 4, 5, 6}.

F A set   H A subset
G A member   J An element

3 List the elements of the set of all the days of the week.

A {Monday, Tuesday, Wednesday, Thursday, Friday}
B {Saturday, Sunday}
C {Sunday, Monday, Tuesday, Wednesday, Thursday, Friday, Saturday}
D {Sunday, Monday, Tuesday, Wednesday, Thursday, Friday}

4 Describe the set {1, 3, 5, 7, 9}.

F Even numbers less than 10
G Prime numbers less than 10
H Numbers less than 10
J Odd numbers less than 10

5 In which sets does the number $^-50$ belong?

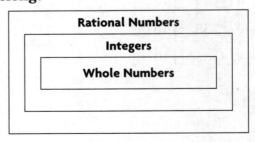

A The sets of whole numbers, integers, and rational numbers
B The sets of integers and rational numbers
C The set of rational numbers
D The set of whole numbers

20

Math Advantage Test Prep

Name _____

Practice Objective 16

DIRECTIONS

Read each question and choose the best answer. Then mark the space for the answer you have chosen.

Use the Venn diagram showing the numbers of students studying a foreign language for the sample question and question 1.

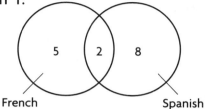

SAMPLE

What does the "2" in the diagram tell you?

A Two students study Spanish.
B Two students study French.
C Two students study both Spanish and French.
D Two students do not study Spanish or French.

1 How many students are studying only Spanish?

A 15 students   C 5 students
B 8 students   D 2 students

2 How many students play both the piano and the organ?

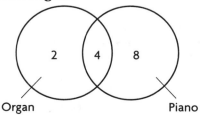

F 12 students   H 8 students
G 10 students   J 4 students

3 How many students play only baseball?

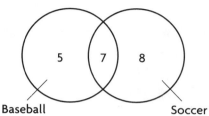

A 5 students   C 8 students
B 7 students   D 12 students

4 How many players are on one or both of these teams?

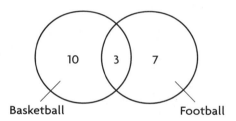

F 7 students   H 8 students
G 12 students   J 20 students

5 Which numbers from the set {1, 3, 4, 5, 6, 8, 9, 11} belong in the shaded section of the diagram?

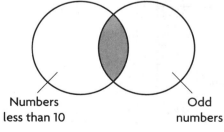

A {11}
B {4, 6, 8}
C {1, 3, 5, 9}
D {1, 3, 4, 5, 6, 8, 9}

Math Advantage Test Prep

Name _____

Practice Objective 17

DIRECTIONS
Read each question and choose the best answer. Then mark the space for the answer you have chosen.

SAMPLE

Kari is shading squares on her hundreds chart according to a number pattern.

[hundreds chart with shaded squares: 6, 15, 24, 33, 42, 51]

Which number should she shade next to continue the pattern?

A 52
B 60
C 61
D 71

1 Minette is shading squares on her hundreds chart according to a number pattern.

[hundreds chart with shaded squares: 4, 15, 26, 37, 48, 59, 70]

Which number should she shade next to continue the pattern?

A 71
B 75
C 80
D 81

2 Carly is working a puzzle that uses this number pattern.

90, 81, 72, 63, ■

What is the missing number?

F 27
G 36
H 45
J 54

3 Denzel is working a puzzle that uses this number pattern.

1.6, 3.3, 5.0, 6.7, ■

What is the missing number?

A 7.4
B 8.4
C 9.4
D 9.7

4 Norm is working a puzzle that uses this number pattern.

$\frac{1}{27}, \frac{1}{9}, \frac{1}{3}, 1,$ ■

What is the missing number?

F 9
G 3
H $2\frac{2}{3}$
J $1\frac{1}{3}$

5 Gavin is working a puzzle that uses this number pattern.

3, 8, 5, 10, 7, 12, ■, ■, ■

What is the last number in this puzzle?

A 18
B 15
C 13
D 11

22   Math Advantage Test Prep

Name _____

Practice Objective 17

DIRECTIONS
Read each question and choose the best answer. Then mark the space for the answer you have chosen.

SAMPLE
How many cubes would you use to make the next figure in this pattern?

A 10 cubes C 14 cubes
B 12 cubes D 18 cubes

1 How many cubes would you use to make the next figure in this pattern?

A 3×2 C 5×2
B 4×2 D 6×2

2 What figure comes next in this pattern?

F

G

H

J

3 What figure comes next in this pattern?

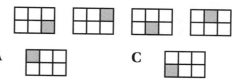

4 The "?" shows where a figure is missing in this pattern. What figure is missing?

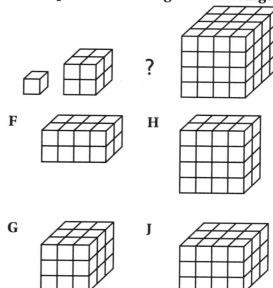

5 How will the number of circles change for the next figure in this pattern?

A Triple   C Increase by 5
B Double   D Increase by 6

Math Advantage Test Prep   23

Name _____

Practice Objective 18

DIRECTIONS
Read each question and choose the best answer. Then mark the space for the answer you have chosen.

SAMPLE

Which property is used in the sentence below?

$2 + (8 + 9) = (2 + 8) + 9$

A Commutative Property
B Associative Property
C Distributive Property
D Identity Property

1 Which property is used in the sentence below?

$7 + 15 = 15 + 7$

A Commutative Property
B Associative Property
C Distributive Property
D Identity Property

2 Which property is used in the sentence below?

$35 + 0 = 35$

F Associative Property
G Commutative Property
H Distributive Property
J Identity Property

3 Which property is used in the sentence below?

$189 \times 1 = 189$

A Commutative Property
B Associative Property
C Distributive Property
D Identity Property

4 Which property is used in the sentence below?

$8 \times 23 = 8 \times (20 + 3)$

F Associative Property
G Commutative Property
H Distributive Property
J Identity Property

5 What inverse operation do you use when solving a division question?

A Addition
B Subtraction
C Multiplication
D Division

Math Advantage Test Prep

Name _____

Practice Objective 18

DIRECTIONS
Read each question and choose the best answer. Then mark the space for the answer you have chosen.

SAMPLE
Which number sentence goes with $56 - n = 16$?

A $56 + n = 16$
B $n + 16 = 56$
C $n \times 16 = 56$
D $n - 16 = 56$

① Which number sentence goes with $35 \div 7 = n$?

A $7 \div n = 35$
B $35 \times 7 = n$
C $n - 7 = 35$
D $n \times 7 = 35$

② Which number sentence goes with $129 + 387 = n$?

F $n + 129 = 387$
G $n - 387 = 129$
H $387 - n = 129$
J $387 \div 3 = n$

③ Which number sentence goes with $30 \times 15 = n$?

A $n \times 15 = 30$
B $n + 15 = 30$
C $30 \div n = 15$
D $n \div 30 = 15$

④ If $25 \times 38 = n \times 25$, then $n =$

F 13
G 38
H 63
J 494

⑤ If $(15 + 6) + 10 = 15 + (n + 10)$, then $n =$

A 6
B 9
C 21
D 31

⑥ If $0.5 \times n = 0.5$, then $n =$

F 1
G 0.5
H 0.25
J 0.1

⑦ If $\frac{11}{15} \times n = 0$, then $n =$

A 1
B $\frac{3}{5}$
C $\frac{4}{15}$
D 0

⑧ If $6 \times 23 = n \times (20 + 3)$, then $n =$

F 120
G 23
H 18
J 6

⑨ If $4 \times 59 = 4 \times (50 + n)$, then $n =$

A 9
B 36
C 200
D 236

⑩ If $7 \times 85 = 7 \times (n + 5)$, then $n =$

F 35
G 56
H 80
J 595

Math Advantage Test Prep

Name _____

> **Practice Objective 19**

DIRECTIONS
Read each question and choose the best answer. Then mark the space for the answer you have chosen.

SAMPLE

If $x + 5 = 8$, then

A $x - (10 - 5) = 8$
B $x - (5 + 8) = 8$
C $x + (10 \div 2) = 8$
D $x + (4 + 4) = 8$

1 If $k + 2 = 10$, then

A $k + (5 + 5) = 10$
B $k + (12 - 10) = 10$
C $k - (10 - 2) = 10$
D $k - (2 + 10) = 10$

2 If $t - 3 = 15$, then

F $t + (5 - 3) = 15$
G $t + (5 - 2) = 15$
H $t - (5 - 2) = 15$
J $t - (15 - 5) = 3$

3 If $30 = h - 10$, then

A $10 = 30 - (h + 5)$
B $10 = h + (17 + 13)$
C $30 = h - (30 + 10)$
D $30 = h + (^-5 + ^-5)$

4 If $9 = 5 + 4$, then

F $9 \div 5 = (5 + 4) - 9$
G $6 + 9 = 6 \times 5 \times 4$
H $9 + 3 = (9 \div 9) \div 3$
J $8 \times 9 = 8 \times (5 + 4)$

5 If $8 = 10 - 2$, then

A $3 \times 8 = 3 \times (2 + 10)$
B $8 + 5 = (10 - 2) + 5$
C $8 \div 4 = (2 + 2) \div 4$
D $8 - 7 = (4 \times 3) - 7$

6 If $6 = 2 + 4$, then

F $6 + 5 = (3 \times 3) + 5$
G $9 - 6 = 9 - (2 + 4)$
H $3 \times 6 = 6 \times (3 \times 3)$
J $6 \div 3 = 6 \div (3 - 1)$

7 If $5 = 3 + 2$, then

A $3 \times 5 = 3 \times (5 - 2)$
B $5 - 4 = (3 + 2) - 4$
C $8 - 5 = 8 + (5 - 1)$
D $15 \div 5 = 15 \div (6 - 3)$

8 If $7 = 9 - 2$, then

F $7 - 4 = (9 - 7) - 4$
G $7 \times 9 = (9 - 7) \times 9$
H $5 + 7 = 5 + (7 - 5)$
J $6 \times 7 = 6 \times (9 - 2)$

Name _____

Practice Objective 20

DIRECTIONS
Read each question and choose the best answer. Then mark the space for the answer you have chosen.

SAMPLE

How many zeros are there in the standard form of 10^6?

A 4 zeros
B 5 zeros
C 6 zeros
D 7 zeros

1 How many zeros are in the standard form of 10^4?

A 3 zeros
B 4 zeros
C 5 zeros
D 6 zeros

2 Which expression represents 5^2?

F $5 + 2$
G 5×2
H 5×5
J $2 \times 2 \times 2 \times 2 \times 2$

3 What is $4 \times 4 \times 4 \times 4 \times 4$ written in exponent form?

A 4^3
B 3^4
C 4^5
D 5^4

4 What is $6 \times 6 \times 6 \times 6$ written in exponent form?

F 6^6
G 6^4
H 4^6
J 4^4

5 What is the value of 3^3?

A 9
B 12
C 27
D 81

6 What is the value of 6^3?

F 216
G 42
H 36
J 18

7 The formula for the area of a square is $A = s^2$. What is the area of a square flower bed whose side measures 2.5 meters?

A 4 square meters
B 5 square meters
C 1.25 square meters
D 6.25 square meters

Math Advantage Test Prep   27

Name _____

Practice Objective 21

DIRECTIONS
Read each question and choose the best answer. Then mark the space for the answer you have chosen.

SAMPLE
Find $5 + 3 \times 2$.

A 10
B 11
C 14
D 16

1 Find $3 \times (9 - 4)$.

A 15
B 23
C 27
D 39

2 Find $20 - (3 + 2)$.

F 25
G 19
H 17
J 15

3 Find $8 + 16 \div 4$.

A 4
B 6
C 12
D 25

4 Find $20 - 15 \div 5$.

F 0
G 1
H 17
J 23

5 Find $15 - 3 + 8 - 2$.

A 18
B 16
C 6
D 2

6 Find $20 - 4^2$.

F 36
G 24
H 12
J 4

7 Find $3 + 2 \times 5 - 1$

A 11
B 12
C 20
D 24

8 Find $6 + 18 \div 3 - 2$.

F 6
G 8
H 10
J 24

9 Find $8 + (10 \div 5) \times 2$.

A 10
B 12
C 20
D 32

10 Find $3^2 + (4 \times 2)$.

F 26
G 17
H 14
J 8

11 Find $3 \times 4 \times (5 - 3)^2$.

A 60
B 57
C 48
D 36

28

Math Advantage Test Prep

Name _____

Practice Objective 22

DIRECTIONS

Read each question and choose the best answer. Then mark the space for the answer you have chosen.

SAMPLE

On a map, which direction is to the left?

A North   C East
B South   D West

Use Shing's map below to help you answer questions 1–5. Each square equals one city block.

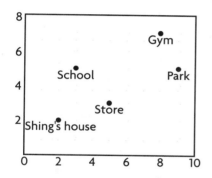
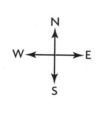

1 Shing left his house and walked 2 blocks north, 7 blocks east, and 1 block north. Where was he?

A Park   C School
B Gym   D Store

2 Jorge left the gym and walked 1 block north, 5 blocks west, and 3 blocks south. Where did he go?

F School   H Gym
G Park   J Shing's house

3 Which location is represented by the ordered pair (9,5)?

A Park.   C School
B Gym   D Store

4 What ordered pair represents the location of the store?

F (10,5)   H (5,3)
G (5,10)   J (3,5)

5 What ordered pair represents the location of the school?

A (6,9)   C (5,3)
B (9.6)   D (3,5)

Use the graph to answer questions 6 and 7.

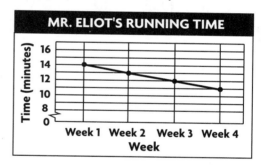

6 During which week was Mr. Eliot's running time 12 minutes?

F Week 1   H Week 3
G Week 2   J Week 4

7 What was Mr. Eliot's running time for Week 1?

A 11 minutes   C 13 minutes
B 12 minutes   D 14 minutes

Math Advantage Test Prep

29

Name _____

Practice Objective 22

DIRECTIONS
Read each question and choose the best answer. Then mark the space for the answer you have chosen.

SAMPLE

How do you locate the point (-4,1) on the coordinate grid? Begin at 0.

A Go up 4 spaces, and go right 1 space.
B Go down 4 spaces, and go right 1 space.
C Go left 4 spaces, and go down 1 space.
D Go left 4 spaces, and go up 1 space.

Use the coordinate plane to help you answer questions 1–2.

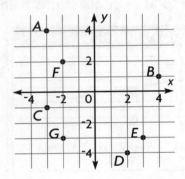

1 Which point is located by the ordered pair (3,-3)?

A Point D
B Point E
C Point F
D Point G

2 Which point is located by the ordered pair (-3,-1)?

F Point A
G Point B
H Point C
J Point D

Use the coordinate plane to help you answer questions 3–6.

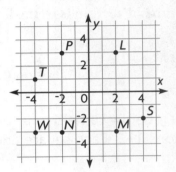

3 Which point is located by the ordered pair (-4,1)?

A Point Q
B Point R
C Point S
D Point T

4 Which point is located by the ordered pair (2,3)?

F Point L
G Point M
H Point N
J Point P

5 What is the ordered pair for point M?

A (2,-3)
B (-2,3)
C (-2,-3)
D (2,3)

6 What is the ordered pair for point P?

F (2,-3)
G (-2,-3)
H (2,3)
J (-2,3)

30

Math Advantage Test Prep

Name _____

Practice Objective 23

DIRECTIONS
Read each question and choose the best answer. Then mark the space for the answer you have chosen.

SAMPLE
A number machine uses the rule "Multiply by 8" to change numbers into other numbers. What is the output?

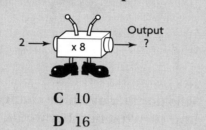

A 4
B 6
C 10
D 16

1 A number machine uses the rule "Add 5" to change numbers into other numbers. What is the output?

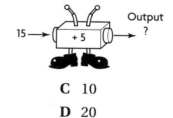

A 3
B 5
C 10
D 20

2 A number machine uses the rule "Subtract 12" to change numbers into other numbers. What is the output?

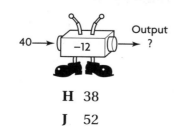

F 4
G 28
H 38
J 52

3 A number machine uses the rule "Divide by 4" to change numbers into other numbers. What is the output?

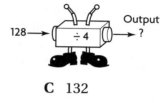

A 32
B 124
C 132
D 512

4 What is the output?

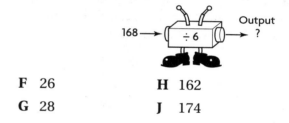

F 26
G 28
H 162
J 174

5 What is the output?

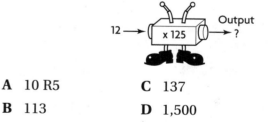

A 10 R5
B 113
C 137
D 1,500

6 What is the output?

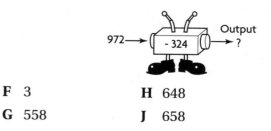

F 3
G 558
H 648
J 658

Math Advantage Test Prep   31

Name _____

Practice Objective 23

DIRECTIONS
Read each question and choose the best answer. Then mark the space for the answer you have chosen.

SAMPLE

A number machine uses a secret rule to change numbers into other numbers.

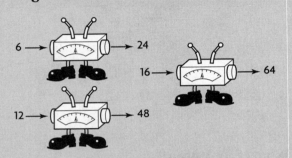

This number machine changed 6 into 24, 12 into 48, and 16 into 64. What number would 8 be changed into?

A 24
B 32
C 48
D 56

1 A number machine uses a secret rule to change numbers into other numbers.

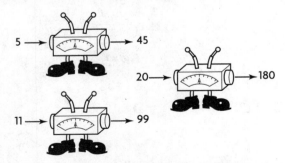

This number machine changed 5 into 45, 11 into 99, and 20 into 180. What number would 7 be changed into?

A 140
B 77
C 63
D 42

2 This number machine changed 9 into 2. What number would 15 be changed into?

F 8
G 24
H 32
J 135

3 This number machine changed 100 into 10. What number would 25 be changed into?

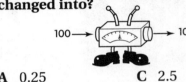

A 0.25
B 2.05
C 2.5
D 250

4 This number machine changed 8 into 20. What number would 4 be changed into?

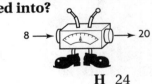

F 9
G 10
H 24
J 29

5 This number machine changed $6\frac{1}{4}$ into $4\frac{1}{2}$. What number would $3\frac{1}{4}$ be changed into?

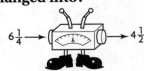

A $1\frac{1}{2}$
B $2\frac{1}{4}$
C 4
D $4\frac{1}{4}$

Math Advantage Test Prep

Name _____

Practice Objective 24

DIRECTIONS
Read each question and choose the best answer. Then mark the space for the answer you have chosen.

SAMPLE

The angle formed by the alligator's mouth is ___?___.

A A right angle   C An obtuse angle
B An acute angle D A straight angle

1 The angle formed by the blades of the scissors is ___?___.

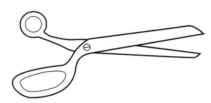

A A right angle   C An obtuse angle
B An acute angle D A straight angle

2 What kind of angle is formed by the two logs?

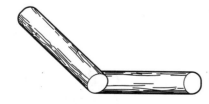

F Right angle   H Obtuse angle
G Acute angle   J Straight angle

3 What kind of angle is formed by two stair steps?

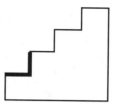

A Right angle   C Obtuse angle
B Acute angle   D Straight angle

4 What kind of angle is formed where the bases of the picture frames meet?

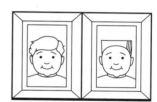

F Right angle   H Obtuse angle
G Acute angle   J Straight angle

5 At which time do the hands of a clock form a straight angle?

A 9:00   C 6:00
B 6:30   D 3:00

6 Barbara is making a scale drawing of her home. What kind of angle will she use when drawing the square corners of the rooms?

F Right angle   H Obtuse angle
G Acute angle   J Straight angle

Math Advantage Test Prep

Name _____

Practice Objective 24

DIRECTIONS
Read each question and choose the best answer. Then mark the space for the answer you have chosen.

SAMPLE
Use a protractor to measure the angle to the nearest degree.

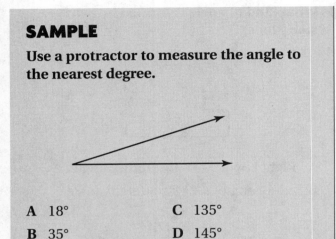

A 18° C 135°
B 35° D 145°

1 Use a protractor to measure the angle to the nearest degree.

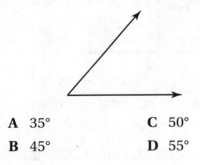

A 35° C 50°
B 45° D 55°

2 Use a protractor to measure the angle to the nearest degree.

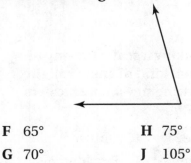

F 65° H 75°
G 70° J 105°

3 Use a protractor to measure the angle to the nearest degree.

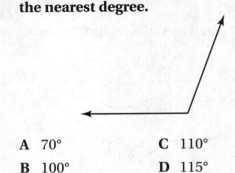

A 70° C 110°
B 100° D 115°

4 Use a protractor to measure the angle to the nearest degree.

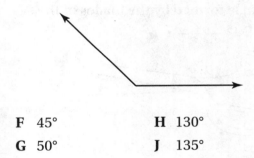

F 45° H 130°
G 50° J 135°

5 Use a protractor to measure the angle to the nearest degree.

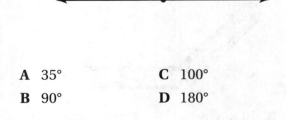

A 35° C 100°
B 90° D 180°

34 Math Advantage Test Prep

Name _____

Practice Objective 25

DIRECTIONS
Read each question and choose the best answer. Then mark the space for the answer you have chosen.

Use the figure below for the sample question and question 1.

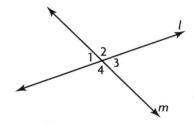

Lines *l* and *m* intersect. Use the figure below to help you answer questions 2–4.

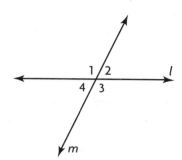

SAMPLE

Which pair of angles are adjacent?

A Angle 1 and Angle 3
B Angle 2 and Angle 4
C Angle 2 and Angle 3
D Line *l* and line *m*

❶ Which pair of angles are vertical angles?

A Angle 1 and Angle 3
B Angle 1 and Angle 2
C Angle 2 and Angle 3
D Angle 3 and Angle 4

❷ Which of these angles and Angle 4 form a pair of adjacent angles?

F Angle 1 only
G Angle 2 only
H Either angle 2 or angle 4
J Either angle 1 or angle 3

❸ Which of these angles and Angle 1 form a pair of vertical angles?

A Angle 1
B Angle 2
C Angle 3
D Angle 4

❹ Which term describes Angle 2 and Angle 4?

F Adjacent angles
G Vertical angles
H Right angles
J Obtuse angles

Math Advantage Test Prep 35

Name _____

Practice Objective 25

DIRECTIONS
Read each question and choose the best answer. Then mark the space for the answer you have chosen.

SAMPLE

___?___ are opposite angles formed when two lines intersect. They are always congruent.

A Congruent line segments
B Perpendicular line segments
C Adjacent angles
D Vertical angles

① ___?___ have the same vertex and a common side.

A Adjacent angles
B Vertical angles
C Right angles
D Congruent line segments

② In the diagram ∠1 and ∠3 are called ___?___.

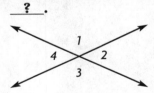

F Acute angles
G Right angles
H Adjacent angles
J Vertical angles

③ In the diagram, ∠2 and ∠3 are called ___?___.

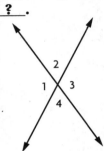

A Congruent angles
B Opposite angles
C Adjacent angles
D Vertical angles

Use the diagram below to help you answer questions 4 and 5.

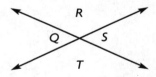

④ In the diagram, ∠T and ∠Q are ___?___.

F Adjacent   H Congruent
G Vertical   J Opposite

⑤ Which of the following choices would not complete the following sentence correctly.

In the diagram, ∠S and ∠Q are ___?___.

A Adjacent   C Congruent
B Vertical   D Opposite

36

Math Advantage Test Prep

Name _____

Practice Objective 26

DIRECTIONS
Read each question and choose the best answer. Then mark the space for the answer you have chosen.

SAMPLE
Which triangle below is a regular polygon?

A Equilateral triangle
B Right triangle
C Acute triangle
D Obtuse triangle

1 What is the name for this polygon?

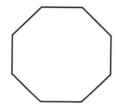

A Pentagon
B Trapezoid
C Hexagon
D Octagon

2 How many sides does a pentagon have?

F 5   H 7
G 6   J 8

For questions 3–4, use the figures below.

Figure A   Figure B   Figure C   Figure D

3 Which figure is a cone?

A Figure A   C Figure C
B Figure B   D Figure D

4 Which figure is a triangular prism?

F Figure A   H Figure C
G Figure B   J Figure D

For questions 5–6, use the figure below.

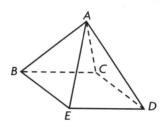

5 How many faces does this solid figure have?

A 4   C 6
B 5   D 7

6 How many edges does this figure have?

F 5   H 7
G 6   J 8

7 The top, bottom, and side views of a solid figure are all squares. What kind of figure is it?

A Hexagonal prism
B Cylinder
C Triangular pyramid
D Cube

Math Advantage Test Prep   37

Name _____

Practice Objective 27

DIRECTIONS

Read each question and choose the best answer. Then mark the space for the answer you have chosen.

SAMPLE

In which figure is the dashed line a line of symmetry?

A 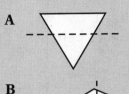   C

B 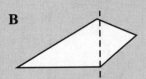   D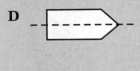

❶ **In which figure is the dashed line a line of symmetry?**

A 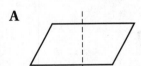   C

B 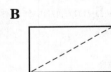   D

❷ **How many lines of symmetry does this figure have?**

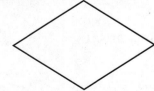

F 0 lines
G 1 line
H 2 lines
J 3 lines

❸ **How many lines of symmetry does this figure have?**

A 6 lines
B 5 lines
C 4 line
D 1 line

❹ **How many lines of symmetry does this figure have?**

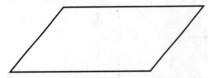

F 0 lines
G 1 line
H 2 lines
J 5 lines

❺ **Toi bought a brownie that was in the shape of a rectangle. She cut the brownie into pieces along its lines of symmetry. How many pieces did she have then?**

A 1 piece
B 2 pieces
C 4 pieces
D 8 pieces

Math Advantage Test Prep

Name _____

DIRECTIONS
Read each question and choose the best answer. Then mark the space for the answer you have chosen.

SAMPLE
Which partial drawing could be completed as a square by drawing exactly one line segment?

A B C D

① Which partial drawing could be completed as a rectangle by drawing exactly one line segment?

A B C D

② Which partial drawing could be completed as a parallelogram by drawing one line segment?

F G H J

③ After drawing the angle below, what kind of triangle could you form by drawing one line segment?

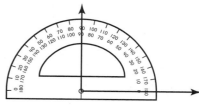

A Acute triangle
B Equilateral triangle
C Obtuse triangle
D Right triangle

④ Two 60° angles were measured on the line segment below. What kind of triangle could you form if you extend the segments until they intersect?

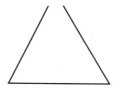

F Right triangle
G Scalene triangle
H Obtuse triangle
J Equilateral triangle

Math Advantage Test Prep 39

Name _____

Practice Objective 29

DIRECTIONS

Read each question and choose the best answer. Then mark the space for the answer you have chosen.

SAMPLE

What construction do the figures below represent?

A Congruent line segments
B Bisected line segments
C Congruent angles
D Perpendicular line segments

1 What construction do the figures below represent?

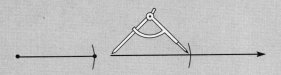

A Perpendicular line segments
B Congruent angles
C Bisected line segments
D Congruent line segments

2 What construction do the figures below represent?

F Bisected line segments
G Parallel line segments
H Congruent angles
J Congruent line segments

3 What construction do the figures below represent?

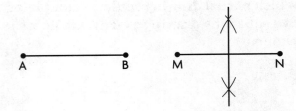

A Perpendicular line segments
B Parallel line segments
C Congruent angles
D Congruent polygons

4 What construction do the figures below represent?

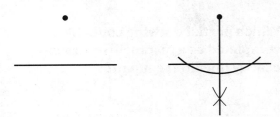

F Congruent angles
G Congruent polygons
H Parallel line segments
J Perpendicular line segments

40

Math Advantage Test Prep

Name _____

DIRECTIONS
Read each question and choose the best answer. Then mark the space for the answer you have chosen.

SAMPLE
Which white figure shows where a gray figure would be if the paper were folded along the heavy dark line?

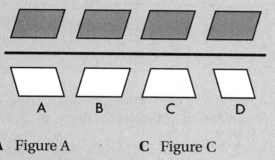

A Figure A
B Figure B
C Figure C
D Figure D

① Which white figure shows where the gray figure would be if the paper were folded along the heavy dark line?

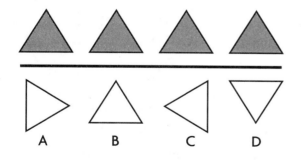

A Figure A
B Figure B
C Figure C
D Figure D

For items 2–4, use triangle *RST*. The coordinates are (1,1), (3,4), and (3,1).

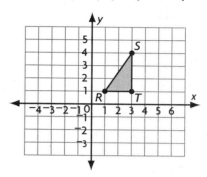

② If you reflect triangle *RST* across the *x*-axis, which coordinates would change?

F Both *x*- and *y*-coordinates
G Only *x*-coordinates
H Only *y*-coordinates
J Neither *x*- nor *y*-coordinates

③ If you reflect triangle *RST* across the *x*-axis, what would the coordinates of the new triangle be?

A $R'(1,-1)$, $S'(3,-3)$ $T'(-3,3)$
B $R'(1,-1)$, $S'(3,-4)$, $T'(3,-1)$
C $R'(1,1)$, $S'(-3,1)$, $T'(-3,4)$
D $R'(-1,-1)$, $S'(-3,-4)$, $T'(-3,-1)$

④ If you reflect triangle *RST* across the *y*-axis, what would the coordinates of the new triangle be?

F $R'''(1,-1)$, $S'''(1,4)$, $T'''(3,-1)$
G $R'''(1,-1)$, $S'''(3,4)$, $T'''(3,-1)$
H $R'''(-1,1)$, $S'''(-1,4)$, $T'''(-3,1)$
J $R'''(-1,1)$, $S'''(-3,4)$, $T'''(-3,1)$

Math Advantage Test Prep

Name _____

Practice Objective 30

DIRECTIONS
Read each question and choose the best answer. Then mark the space for the answer you have chosen.

SAMPLE
What moves were made to transform figure 1 into each new position?

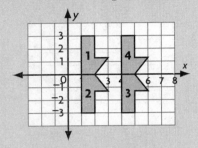

A Rotation, reflection, rotation
B Translation, reflection, translation
C Reflection, translation, reflection
D Rotation, translation, reflection

1 What moves were made to transform figure 1 into each new position?

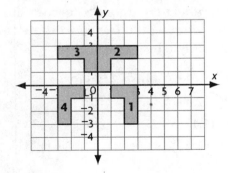

A Rotation, reflection, rotation
B Reflection, translation, reflection
C Translation, reflection, translation
D Rotation, translation, reflection

2 What moves were made to transform figure 1 into each new position?

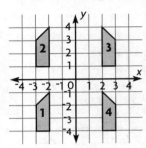

F Reflection, translation, reflection
G Translation, reflection, reflection
H Translation, rotation, translation
J Translation, reflection, translation

3 What moves were made to transform figure 1 into each new position?

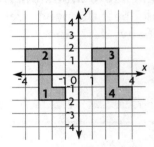

A Rotation, translation, rotation
B Translation, reflection, translation
C Rotation, reflection, rotation
D Reflection, translation, reflection

Math Advantage Test Prep

Name _____

Practice Objective 31

DIRECTIONS
Read each question and choose the best answer. Then mark the space for the answer you have chosen.

SAMPLE

If *C* is the center of the circle, which line segment is a radius of the circle?

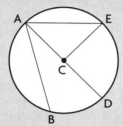

A $\overline{AB}$ C $\overline{AD}$
B $\overline{CA}$ D $\overline{AE}$

1 If *C* is the center of the circle, which line segment is a radius of the circle?

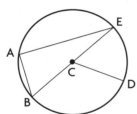

A $\overline{AB}$ C $\overline{BE}$
B $\overline{AE}$ D $\overline{CD}$

2 If *C* is the center of the circle, which line segment is a diameter of the circle?

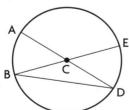

F $\overline{CA}$ H $\overline{BD}$
G $\overline{CB}$ J $\overline{BE}$

3 If *C* is the center of the circle, which line segment is a diameter of the circle?

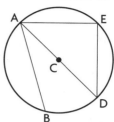

A $\overline{AB}$ C $\overline{AE}$
B $\overline{AD}$ D $\overline{DE}$

4 If *C* is the center of the circle, which line segment is a chord of the circle?

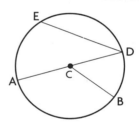

F $\overline{CD}$ H $\overline{CA}$
G $\overline{CB}$ J $\overline{ED}$

5 The diameter of a pie measures 10 inches. If the pie is cut into six congruent pieces, what is the length of each cut side of one piece of pie?

A 2 inches
B 5 inches
C 10 inches
D 15 inches

Math Advantage Test Prep   43

Name _____

Practice Objective 32

DIRECTIONS

Read each question and choose the best answer. Then mark the space for the answer you have chosen.

SAMPLE

Use your centimeter ruler to measure the length of the path from *A* to *B*.

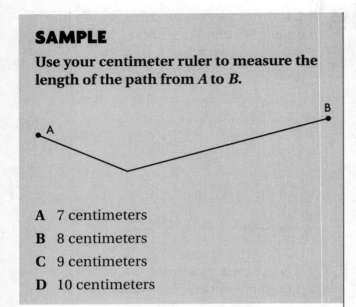

A 7 centimeters
B 8 centimeters
C 9 centimeters
D 10 centimeters

① Use your centimeter ruler to measure the perimeter of the figure below.

A 2 cm   C 15 cm
B 8 cm   D 16 cm

② What is the area of the figure in item 1?

F 8 cm²   H 16 cm²
G 15 cm²   J 32 cm²

③ Miss Que has to go from the school to the grocery store and the library before going to a music lesson. How much difference is there between the routes *SGLM* and *SLGM*?

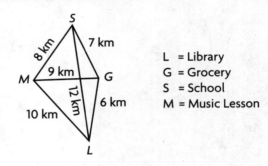

L = Library
G = Grocery
S = School
M = Music Lesson

A SLGM is 1 km shorter.
B SLGM is 2 km shorter.
C SGLM is 4 km shorter.
D SGLM is 1 km shorter.

④ If Miss Que chooses the shorter route in item 3, how many kilometers will she travel?

F 21 km   H 25 km
G 23 km   J 27 km

⑤ If Miss Que chooses the longer route in item 3, how many kilometers will she travel?

A 21 km   C 25 km
B 23 km   D 27 km

44   **Math Advantage Test Prep**

Name _____

Practice Objective 32

DIRECTIONS
Read each question and choose the best answer. Then mark the space for the answer you have chosen.

SAMPLE

A passenger jet travels 550 miles per hour. How far would the plane travel in 3.5 hours?

A 1,500 miles
B 1,550 miles
C 1,650 miles
D 1,925 miles

1 A high-speed train travels 300 kilometers per hour. How far would the train travel in 2.5 hours?

A 750 km
B 600 km
C 450 km
D 120 km

2 Columbus is 90 miles from Louisville. How long will it take a car traveling 60 miles per hour to go from Columbus to Louisville?

F $\frac{1}{2}$ hour
G $\frac{2}{3}$ hour
H $1\frac{1}{2}$ hours
J 3 hours

3 Firewood is sold by the cord. A cord is a stack of wood that is 8 feet long by 4 feet wide by 4 feet high. What is the volume of a cord of wood?

A 32 ft^3
B 128 ft^3
C 64 ft^3
D 320 ft^3

4 If the length of each side of the cube doubles, how does its volume change?

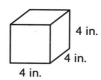

F It is twice the original cube.
G It is 4 times the original cube.
H It is 8 times the original cube.
J It is 16 times the original cube.

5 How can you change this box so that it will hold twice as much?

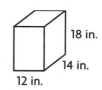

A Double only one dimension.
B Double only two dimensions.
C Double all three dimensions.
D Add 2 in. to each dimension.

6 Emily is planning to paint just the sides of a large block that is 13 cm long, 11 cm wide, and 9 cm high. How much surface area will she paint?

F 143 cm^2
G 432 cm^2
H 484 cm^2
J 620 cm^2

Math Advantage Test Prep 45

Name _____

Practice Objective 33

DIRECTIONS
Read each question and choose the most appropriate measure.
Then mark the space for the answer you have chosen.

SAMPLE
The length of a pen

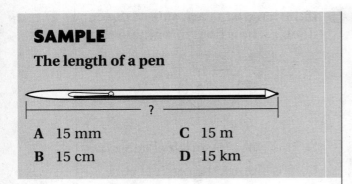

A 15 mm   C 15 m
B 15 cm   D 15 km

1 The length of a spoon

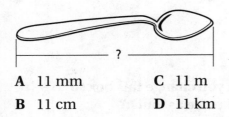

A 11 mm   C 11 m
B 11 cm   D 11 km

2 The distance between two airports

F 35 km   H 35 cm
G 35 m   J 35 mm

3 The length of a bike trail

A 100 m   C 1,000 cm
B 10 km   D 20 m

4 The length of a crayon

F 8 mm   H 8 m
G 8 cm   J 8 km

5 The diameter of a nickel

A 2 cm   C 2 m
B 2 mm   D 2 km

6 The distance across a bridge

F 125 cm   H 2 km
G 200 mm   J 200 km

7 The length of a paper clip

A 3 mm   C 3 m
B 3 cm   D 3 km

8 The width of your math book

F 22 km   H 22 cm
G 22 m   J 22 mm

9 Each of four friends measured the length of Beth's house. Which of their measurements was more precise?

A 62 m   C 610 dm
B 60 m   D 6,159 cm

Math Advantage Test Prep

Name _____

Practice Objective 33

DIRECTIONS
Read each question and choose the most appropriate measure.
Then mark the space for the answer you have chosen.

SAMPLE
The capacity of a juice glass

A 100 mL C 10 L
B 10 mL   D 125 L

① The capacity of a pitcher of punch

A 2 mL   C 2 L
B 20 mL D 20 L

② The capacity of a paint can

F 4 mL   H 4 L
G 400 mL J 40 L

③ The capacity of a drinking glass

A 300 L   C 3 L
B 300 mL D 30 mL

④ The capacity of an auto gas tank

F 5 L   H 50 L
G 50 mL   J 500 L

⑤ The mass of a bowling ball

A 8 kg   C 800 g
B 8 g   D 80 mg

⑥ The mass of an egg

F 500 mg   H 50 g
G 5 mg   J 5 kg

⑦ The mass of a jar of jam

A 90 kg   C 900 mg
B 900 g   D 9 kg

⑧ The mass of a paper clip

F 20 kg   H 2 g
G 2 kg   J 20 mg

⑨ The mass of stick of gum

A 4 mg   C 4 kg
B 4 g   D 40 kg

⑩ Each of four basketball players measured the mass of a basketball. Which of their measurements was more precise?

F 0.6 kg   H 0.55 kg
G 0.5 kg   J 566 g

⑪ Each of four cooks measured the capacity of a pitcher. Which of their measurements was more precise?

A 4.5 L   C 4.25 L
B 4 L   D 4,259 mL

Math Advantage Test Prep 47

Name _____

Practice Objective 33

DIRECTIONS
Read each question and choose the most appropriate measure.
Then mark the space for the answer you have chosen.

SAMPLE
The width of a school locker
A 14 in.   C 11 ft
B 42 in.   D $2\frac{1}{2}$ yd

1 The height of a toaster
A 7 ft   C 1 yd
B 7 in.   D 2 ft

2 The height of a doorknob
F 20 in.   H 5 ft
G 3 ft   J 2 yd

3 The height of a woman
A 12 ft   C 100 in.
B 1 yd   D 5 ft

4 The distance from Los Angeles to Atlanta
F 2,200 in.   H 2,200 mi
G 5,000 ft   J 5,000 mi

5 The height of a desk
A 14 in.   C 20 ft
B $2\frac{1}{2}$ ft   D 14 yd

6 The length of a tube of toothpaste
F 75 in.   H 6 in.
G 2 ft   J 2 yd

7 The length of a river
A 300 mi   C 400 yd
B 1,000 in.   D 150 ft

8 The length of a toothpick
F 2 ft   H 20 in.
G 40 yd   J 2 in.

9 The length of a driveway from the curb to the garage
A 17 in.   C 1,700 ft
B 17 yd   D 170 mi

10 The length of a basketball court
F 28 in.   H 28 yd
G 28 ft   J 2.8 mi

11 Each of four students measured the length of their classroom. Which of their measurements was more precise?
A 4 yd   C 3 yd
B 3.5 yd   D $11\frac{1}{2}$ ft

Math Advantage Test Prep

Name _____

Practice Objective 33

DIRECTIONS
Read each question and choose the most appropriate measure.
Then mark the space for the answer you have chosen.

SAMPLE
The weight of a turkey

A 12 lb   C 45 lb
B 20 oz   D 60 oz

1 The weight of a sack of potatoes

A 2 oz   C 5 lb
B 25 oz   D 500 lb

2 The weight of a bicycle

F 200 lb   H 2 lb
G 20 lb   J 20 oz

3 The weight of a pumpkin

A 1 ton   C 200 lb
B 12 lb   D 20 oz

4 The weight of a truck

F 10 T   H 1,000 oz
G 100 lb   J 10 lb

5 The capacity of a coffee mug

A 9 fl oz   C 2 pt
B 30 fl oz D 10 cups

6 The capacity of a vegetable serving bowl

F 1 gal   H 3 pt
G 10 qt   J $\frac{1}{2}$ cup

7 The capacity of a juice box

A 8 pt   C 8 fl oz
B 1 qt   D 1 fl oz

8 The capacity of a milk pitcher

F 10 qt   H 20 fl oz
G 20 qt   J 2 fl oz

9 The capacity of a kitchen sink

A 50 qt   C 10 cups
B 4 pt   D 1 gal

10 The capacity of a can of soda

F 2 cups   H 1.75 gal
G 2.5 qt   J 3 gal

11 Enid and three friends measured the weight of a cocker spaniel. Which of their measurements was more precise?

A 672 oz   C 42 lb
B 672.5 oz   D 42.2 lb

Math Advantage Test Prep

Name _____

Practice Objective 34

DIRECTIONS
Read each question and choose the best answer. Then mark the space for the answer you have chosen. If a correct answer is *not here,* mark the space for NH.

SAMPLE
150 cm =
- A 15 m
- B 0.15 mm
- C 1.5 m
- D 1.5 km
- E NH

1 62 cm =
- A 620 mm
- B 6,200 m
- C 6.2 km
- D 0.062 m
- E NH

2 158 cm =
- F 1.58 mm
- G 1.58 m
- H 0.158 m
- J 1.58 km
- K NH

3 34 mm =
- A 3.4 cm
- B 3.4 m
- C 0.34 m
- D 0.34 km
- E NH

4 2,340 m =
- F 0.234 km
- G 2.34 km
- H 23.4 km
- J 234 km
- K NH

5 8,070 mL =
- A 8.07 mL
- B 807 L
- C 80,700 mL
- D 80.7 L
- E NH

6 6.1 L =
- F 61 mL
- G 610 mL
- H 6,100 mL
- J 61,000 mL
- K NH

7 7.6 kg =
- A 7,600 g
- B 760 g
- C 7,600 mg
- D 760 mg
- E NH

8 2.8 g =
- F 2,800 kg
- G 28 kg
- H 280 mg
- J 2,800 mg
- K NH

9 A can of beans weighs 425 grams. How many kilograms will 24 cans weigh?
- A 4.25 kg
- B 42.5 kg
- C 10.2 kg
- D 10,200 kg
- E NH

50

Math Advantage Test Prep

Name _____

DIRECTIONS
Read each question and choose the best answer. Then mark the space for the answer you have chosen. If a correct answer is *not here,* mark the space for NH.

SAMPLE
144 in. =

A 12 yd
B 6 yd
C 12 ft
D 6 ft
E NH

1 10 mi =

A 5,280 ft
B 10,000 ft
C 52,800 ft
D 528,000 ft
E NH

2 1,760 yd =

F 5,280 ft
G 5,280 in.
H 176 ft
J $146\frac{2}{3}$ ft
K NH

3 252 in. =

A 9,072 yd
B 3,024 ft
C 21 ft
D 7 ft
E NH

4 51 ft =

F 612 yd
G 153 yd
H 612 in.
J 17 in.
K NH

5 10 qt =

A 20 gal
B 5 gal
C 20 c
D 5 pt
E NH

6 28 qt =

F 7 gal
G 4 gal
H 14 pt
J 56 c
K NH

7 16 gal =

A 32 qt
B 32 pt
C 64 qt
D 264 pt
E NH

8 8 lb =

F 8 oz
G 64 oz
H 96 oz
J 128 oz
K NH

9 Tasha wants to make 160 cups of punch for her party. How many quarts of punch will she make?

A 40 qt
B 80 qt
C 320 qt
D 640 qt
E NH

Math Advantage Test Prep 51

Name _____

Practice Objective 35

DIRECTIONS

Read each question and choose the best approximate answer. Then mark the space for the answer you have chosen. If a correct answer is *not here*, mark the space for NH.

SAMPLE
80 in. ≈ ■ m

A 20
B 7
C 2
D 1
E NH

1 5 m ≈ ■ yd

A 2
B 5
C 15
D 500
E NH

2 9 ft ≈ ■ m

F 90
G 27
H 3
J 2
K NH

3 2 mi ≈ ■ km

A 1
B 4
C 8
D 20
E NH

4 $\frac{1}{2}$ in. ≈ ■ cm

F 12
G 4
H 2
J 1
K NH

5 8 yd ≈ ■ m

A $\frac{2}{3}$
B 16
C 24
D 96
E NH

6 6 qt ≈ ■ L

F 12
G 6
H 3
J 1.5
K NH

7 1 gal ≈ ■ L

A 16
B 12
C 8
D 4
E NH

8 3 c ≈ ■ mL

F 75
G 750
H 900
J 1,500
K NH

9 A specialty can of coffee weighs 2 pounds. About how many kilograms does it weigh?

A About 1 kg
B About 2.5 kg
C About 3 kg
D About 4 kg

Math Advantage Test Prep

Name _____

Practice Objective 36

DIRECTIONS
Read each question and choose the best answer. Then mark the space for the answer you have chosen.

SAMPLE
Estimate the area. Each square is 1 square centimeter (cm^2).

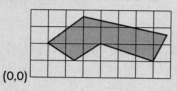
(0,0)

A About 4 cm^2
B About 8 cm^2
C About 12 cm^2
D About 21 cm^2

Use the figure for questions 1 and 2. Each square is 1 square centimeter.

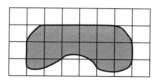

1 Estimate the perimeter.

A About 20 cm   C About 14 cm
B About 18 cm   D About 10 cm

2 Estimate the area.

F About 13 cm^2 H About 20 cm^2
G About 16 cm^2 J About 25 cm^2

Use the figure for questions 3 and 4. Each square is 1 cm^2.

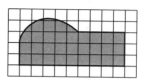

3 Estimate the perimeter.

A About 15 cm   C About 24 cm
B About 20 cm   D About 30 cm

4 Estimate the area.

F About 26 cm^2 H About 33 cm^2
G About 30 cm^2 J About 45 cm^2

Alfonso drew this design for art class. Use the design for questions 5 and 6. Each square is 1 cm^2.

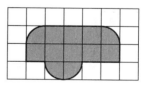

5 Estimate the perimeter.

A About 8 cm   C About 16 cm
B About 12 cm   D About 25 cm

6 Estimate the area.

F About 15 cm^2 H About 11 cm^2
G About 14 cm^2 J About 9 cm^2

Math Advantage Test Prep   53

Name _____

Practice Objective 37

DIRECTIONS
Read each question and choose the best answer. Then mark the space for the answer you have chosen.

Use your centimeter ruler and this map to help you answer the sample question and question 1.

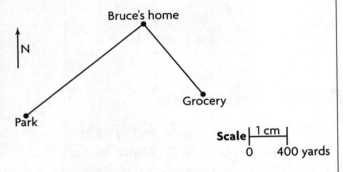

SAMPLE
What is the actual distance from Bruce's home to the park?

A 1,600 yd   C 800 yd
B 1,200 yd   D 400 yd

1 What is the actual distance from Bruce's home to the grocery?

A 400 yd   C 800 yd
B 600 yd   D 1,000 yd

2 The distance from Tuan's city to a neighboring city is 4 inches on a map. What is the actual distance if the scale is 1 in.:20 mi?

F 80 mi   H 50 mi
G 60 mi   J 5 mi

3 The distance from Mel's home to Hanover measures $5\frac{1}{2}$ inches on a scale drawing. What is the actual distance if the scale is 1 in.:100mi?

A 505 mi   C 555 mi
B 550 mi   D 650 mi

4 A tiny screw used in eyeglass frames measures 4 millimeters in length. What is the length of the screw in a scale drawing if the scale is 2 cm:1 mm?

F 4 cm   H 8 cm
G 6 cm   J 10 cm

5 The actual length of a bird feeder is 10 inches. What is the length of the bird feeder in a scale drawing if the scale is 1 in.:4 in.?

A $1\frac{1}{2}$ in.   C 5 in.
B $2\frac{1}{2}$ in.   D 10 in.

6 A school bus made a round trip from Chester to Richmond for a soccer game. The map distance between Chester and Richmond is 4 in. What is the actual round-trip distance if the scale is 1 in. = 3 mi?

F 12 mi   H 24 mi
G 18 mi   J 36 mi

54

Math Advantage Test Prep

Name _____

DIRECTIONS
Read each question and choose the best answer. Then mark the space for the answer you have chosen.

SAMPLE

The length-to-width ratio of a scale drawing is ___?___ the length-to-width ratio of the original object.

A Larger than
B Smaller than
C The same as
D Twice

1 Which of the following would be the best scale to use if you are making a scale drawing of a new car?

A 5 m:1 cm
B 1 in.:2 ft
C 2 ft:1 in.
D 1 in.:12 ft

2 Jamie wants to make a scale drawing of a beetle with an actual length of 3 centimeters. She decides to use 12 centimeters for the length in the scale drawing. What scale should she use?

F 1 in.:12 in.
G 1 in.:4 in.
H 1 cm:4 cm
J 4 cm:1 cm

3 Arafat uses 10 centimeters to represent the width of a flower in a scale drawing. The actual width is 2 centimeters. What scale should he use?

A 1 cm:10 cm   C 2 cm:10 cm
B 5 cm:1 cm   D 5 cm:2 cm

4 Maribelle decides to use 1 cm:4 m for a scale drawing of a house. The actual length is 20 meters. How long should Maribelle make the length of the house in the scale drawing?

F 5 centimeters   H 10 centimeters
G 8 centimeters   J 15 centimeters

5 Ken decides to use a scale of 5 in.:2 in. for a scale drawing of a beetle. The actual length of the beetle is 3 inches. How long should Ken make the length in the drawing?

A 6 inches   C $7\frac{1}{2}$ inches
B 7 inches   D 15 inches

6 Yasmeen makes a map using a scale of 1 in.:50 mi. If one of the map distances is 4 inches, what does n represent in the proportion $\frac{1}{50} = \frac{4}{n}$?

F The actual distance in miles
G The map distance in inches
H The straight-line distance in inches
J The scale

Math Advantage Test Prep   55

Name _____

Practice Objective 38

DIRECTIONS
Read each question and choose the best answer. Then mark the space for the answer you have chosen.

SAMPLE
Use your centimeter ruler to help you answer the question. What is the perimeter of this triangular sticker?

A 10.5 cm
B 7.5 cm
C 7 cm
D 6.5 cm

1 Use your centimeter ruler to help you answer the question. What is the perimeter of this rectangular name tag?

A 8.5 cm
B 11 cm
C 16 cm
D 17 cm

2 The city's rectangular parking lot is 120 yd long and 90 yd wide. How much fencing is needed to enclose the parking lot?

F 210 yd
G 420 yd
H 480 yd
J 840 yd

3 Max put up a fence around this garden to keep small animals out. What is the perimeter of this vegetable garden?

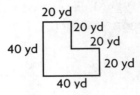

A 120 yd
B 140 yd
C 160 yd
D 180 yd

4 The perimeter of this town is 103 miles. What is the missing length?

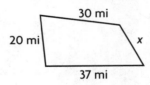

F 26 mi
G 22 mi
H 20 mi
J 16 mi

5 A puzzle is in the shape of a regular hexagon. The perimeter of the puzzle is 48 in. What is the length of each side?

A 6 in.
B 8 in.
C 12 in.
D 16 in.

6 The perimeter of a rectangular door is 20 ft. The door is 3 ft wide. How tall is the door?

F 10 ft
G 8.5 ft
H 7 ft
J 3 ft

56 Math Advantage Test Prep

Name _____

Practice Objective 38

DIRECTIONS
Read each question and choose the best answer. Then mark the space for the answer you have chosen.

SAMPLE
Use your centimeter ruler to help answer this question. Which stamp has an area of 4 cm²?

A    C

B   D

① Use your centimeter ruler to help answer this question. Which eraser has an area of 3 cm²?

A    C

B 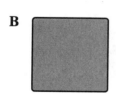   D

② Marcus drew a square that measured 5 inches on each side. If he doubles the length of each side, what will the area of the square be?

F 10 in.²   H 25 in.²
G 20 in.²   J 100 in.²

③ Leroy has a rectangular garden that is 20 feet by 12 feet. He decides to double the garden's area. Which of the following dimensions will double the area of his garden?

A 30 ft by 12 ft   C 40 ft by 24 ft
B 40 ft by 12 ft   D 80 ft by 24 ft

④ Lena built a square table that measured 4 feet on each side. She used 16 square tiles to cover the top of the table. She then built another square table that measured 2 feet on each side. If she uses the same kind of tiles to cover the top of the table, how many tiles will she need?

F 14 tiles   H 8 tiles
G 12 tiles   J 4 tiles

⑤ Corey is constructing a 12-foot by 3-foot rectangular walkway made of red bricks. If each brick has an area of 1 square foot and costs $3.50, what is the total cost of the bricks?

A $105.00   C $126.00
B $108.00   D $133.00

⑥ Kendra bought 36 carpet tiles to carpet her rectangular office. Michaelene's office is twice as long and twice as wide as Kendra's office. How many carpet tiles will be needed to carpet Michaelene's office?

F 18 tiles   H 90 tiles
G 72 tiles   J 144 tiles

Math Advantage Test Prep

Name _____

Practice Objective 39

DIRECTIONS
Read each question and choose the best answer. Then mark the space for the answer you have chosen.

SAMPLE

Find the perimeter of the square. Use the formula $P = 4s$.

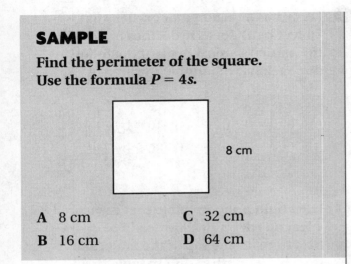

A 8 cm
B 16 cm
C 32 cm
D 64 cm

1 Find the perimeter of the square. Use the formula $P = 4s$.

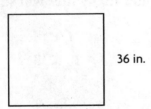

36 in.

A 9 in.
B 72 in.
C 144 in.
D 1,296 in.

2 Find the perimeter of the rectangle. Use the formula $P = (2 \times l) + (2 \times w)$.

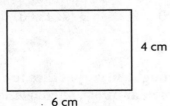

4 cm
6 cm

F 10 cm
G 20 cm
H 24 cm
J 48 cm

3 Find the perimeter of the regular hexagon. Use the formula $P = 6s$.

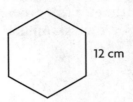

12 cm

A 48 cm
B 72 cm
C 82 cm
D 144 cm

4 Find the circumference of the circle. Use the formula $C = \pi d$. Use 3.14 for π.

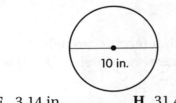

10 in.

F 3.14 in.
G 30.14 in.
H 31.4 in.
J 314 in.

5 Find the circumference of the circle. Use the formula $C = \pi d$. Use $\frac{22}{7}$ for π.

7 in.

A 14 in.
B 22 in.
C 44 in.
D 88 in.

Math Advantage Test Prep

Name _____

Practice Objective 39

DIRECTIONS
Read each question and choose the best answer. Then mark the space for the answer you have chosen.

SAMPLE
Find the area of the square.
Use the formula $A = s^2$.

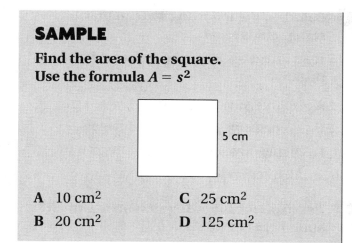

A 10 cm² C 25 cm²
B 20 cm² D 125 cm²

1 Find the area of the rectangle.
Use the formula $A = lw$.

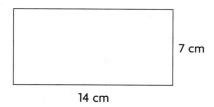

A 196 cm²   C 42 cm²
B 98 cm²   D 21 cm²

2 Find the area of the parallelogram.
Use the formula $A = bh$.

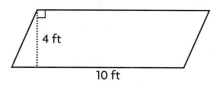

F 20 ft²   H 40 ft²
G 28 ft²   J 80 ft²

3 Find the area of the circle. Use the formula $A = \pi r^2$. Use 3.14 for π. Round the answer to the nearest whole unit.

A About 36 m²   C About 113 m²
B About 108 m²   D About 144 m²

4 Find the volume of the prism.
Use the formula $V = lwh$.

F 44 in.³   H 1,512 in.³
G 186 in.³   J 3,024 in.³

5 Find the volume of the cylinder. Use the formula $V = \pi r^2 h$. Use 3.14 for π. Round the answer to the nearest whole unit.

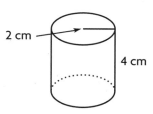

A 16 cm³   C 32 cm³
B 25 cm³   D 50 cm³

Math Advantage Test Prep   59

Name _____

Practice Objective 40

DIRECTIONS
Read each question and choose the best answer. Then mark the space for the answer you have chosen.

SAMPLE

Based on the following statement, which statement is true?

If a number has two digits, then it is less than 100.

A All numbers less than 100 have two digits.
B No numbers less than 100 have two digits.
C All two-digit numbers are less than 100.
D No two-digit numbers are less than 100.

1 Based on the following statement, which statement is true?

If an angle measures 50°, then it is an acute angle.

A All acute angles measure 50°.
B No acute angles measure 50°.
C All angles that measure 50° are acute.
D No 50° angles are acute.

2 Based on the following statement, which statement is true?

If a figure is a square, then it is a quadrilateral.

F All squares are quadrilaterals.
G No squares are quadrilaterals.
H All quadrilaterals are squares.
J No quadrilaterals are squares.

3 Based on the following statement, which statement is true?

If an integer is less than 0, then it is negative.

A No integers are negative.
B No negative numbers are integers.
C All negative numbers are integers.
D All integers less than 0 are negative.

4 Based on the following statement, which statement is true?

If a figure is a rectangle, then it is a parallelogram.

F All parallelograms are rectangles.
G No parallelograms are rectangles.
H All rectangles are parallelograms.
J No rectangles are parallelograms.

5 Based on the following statement, which statement is true?

If two lines are perpendicular, then they intersect.

A All intersecting lines are perpendicular.
B Two perpendicular lines intersect.
C No intersecting lines are perpendicular.
D No perpendicular lines intersect.

6 Based on the following statement, which statement is true?

If a figure is a square, then it it a rectangle.

F All rectangles are squares.
G All squares are rectangles.
H No squares are rectangles.
J No rectangles are squares.

Math Advantage Test Prep

Name _____

DIRECTIONS
Read each question and choose the best answer. Then mark the space for the answer you have chosen.

For the sample question and question 1, use the stem-and-leaf plot of test scores for a sixth-grade class.

Class Test Scores

Stem	Leaves
5	8 9
6	0 4 7 8 8
7	0 1 3 4 6 7 9 9
8	0 0 1 2 2 4 5 6 6 8 9
9	1 3 5 6 7

SAMPLE
What score is shown by the fourth stem and its second leaf?

A 68
B 79
C 80
D 81

1 Where should a score of 90 be placed on the stem-and-leaf plot?

A The fourth stem and its eleventh leaf
B The fifth stem and its first leaf
C The fifth stem and its third leaf
D The fifth stem and its fifth leaf

2 The histogram below shows the ages of people at a movie. What label is missing from the histogram?

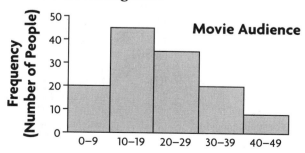

F Frequency
G Number of people
H Audience
J Age

3 What would the histogram for the frequency table below look like?

15-Mile Race	
Minutes	Runners
0 - 59	20
60 - 119	35
120 - 179	10
180 - 239	5

A
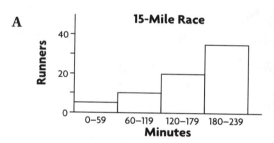

B
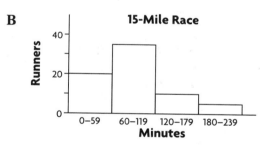

C
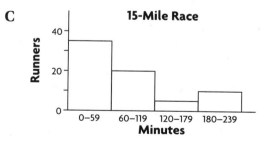

D
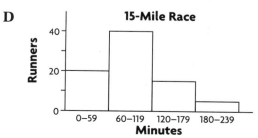

Math Advantage Test Prep

Name _____

Practice Objective 41

DIRECTIONS
Read each question and choose the best answer. Then mark the space for the answer you have chosen.

For the sample question and question 1, use the table showing the results when students were surveyed about their favorite pizzas.

Favorite Pizza	
Plain	15
Mushrooms	35
Pepperoni	50

SAMPLE
How many students were surveyed?

A 50 students
B 85 students
C 90 students
D 100 students

① Barry is using the data about favorite pizzas to make a circle graph. What will Barry's graph look like?

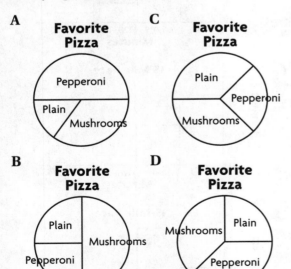

② The Schools in Northwest County conducted a survey of 100 students to find out if they liked the idea of changing the school schedule. What did the schools learn from the survey?

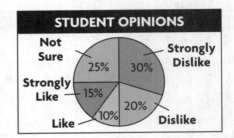

F Of the students surveyed, 25% favored the idea.
G Of the students surveyed, 60% disliked the idea.
H Opinions were evenly divided.
J More than half of the students surveyed were against the idea.

③ Which statement about the 8th-grade band members is true?

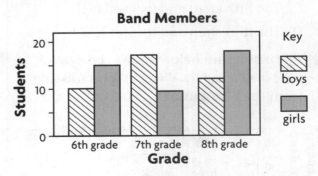

A There are more girls than boys.
B There are more boys than girls.
C There are fewer 8th graders than 7th graders.
D There are fewer 8th graders than 6th graders.

62 **Math Advantage Test Prep**

Name _____

Practice Objective 41

DIRECTIONS
Read each question and choose the best answer. Then mark the space for the answer you have chosen.

For the sample question and questions 1 and 2, use the graph below.

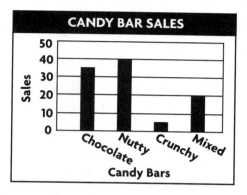

SAMPLE

Which kind of candy bar was the most popular?

A Chocolate
B Nutty
C Crunchy
D Mixed

1 How does the number of mixed candy bars sold compare with the number of crunchy candy bars sold?

A The same number of each was sold.
B Twice as many crunchy candy bars were sold.
C Twice as many mixed candy bars were sold.
D About four times as many mixed candy bars were sold.

2 The students plan to sell a fruit and nut candy bar at their next sale. They still want only four kinds of candy bars. Which candy bar should they stop selling?

F Chocolate
G Nutty
H Crunchy
J Mixed

3 What was the trend for sales of cars?

A Increased
B Decreased
C Stayed the same
D Less than truck sales

4 The enrollment at Shields Middle School was 105 in 1994, 110 in 1995, 115 in 1996, 120 in 1997, and 125 in 1998. If this data is graphed, what trend for enrollment would it show?

F Decreasing
G Increasing
H Staying the same
J Not enough data

5 According to the graph below, about how many times greater are the sales in Week 4 than the sales in Week 1?

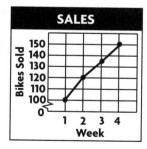

A About $1\frac{1}{2}$ times greater
B About twice as great
C About three times as great
D About five times as great

Math Advantage Test Prep

Name _____

Practice Objective 42

DIRECTIONS
Read each question and choose the best answer. Then mark the space for the answer you have chosen.

SAMPLE
What is the mean (average) for this set of data?
20, 24, 27, 1, 23

A 18   C 20
B 19   D 26

❹ What is the median for this set of data?
49, 52, 58, 49, 56

F 58   H 52
G 52.8   J 49

❶ What is the mean (average) for this set of data?
51, 48, 39, 53, 47, 44, 42, 55, 44

A 46   C 47
B 46.5   D 47.5

❺ What is the median for this set of data?
45, 58, 37, 29, 76, 19, 68, 91, 57

A 45   C 54.2
B 47   D 57

❷ What is the mode for this set of data?
14, 21, 17, 25, 20, 17

F 17   H 18.5
G 18   J 19

❻ What is the range for this set of data?
34, 47, 38, 42, 15, 59, 71

F 15   H 56
G 50   J 71

❸ What is the mode for this set of data?
11, 16, 14, 18, 18, 13

A 11   C 15
B 16   D 18

❼ What is the range for this set of data?
11, 15, 13, 12, 15, 13, 12, 14,
13, 15, 15, 11, 16, 12, 11, 12

A 5   C 11
B 6   D 12

64

Math Advantage Test Prep

Name _____

Practice Objective 42

DIRECTIONS

Read each question and choose the best answer. Then mark the space for the answer you have chosen.

SAMPLE

A figure skater was awarded these scores by six judges: 7.5, 8.5, 8.0, 8.2, 8.3, and 7.5. What is the mean (average) score?

A 8.5
B 8.1
C 8.0
D 7.5

For questions 4 and 5, use the line plot. It shows scores on a math test taken by 19 students.

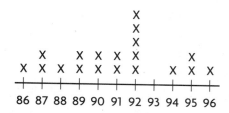
Test Scores on a Math Test

① Dan scored these points during the basketball season: 14, 21, 17, 25, 17, 18, 16, 22, 17, and 23. What is the mean number of points Dan scored?

A 11
B 17
C 19
D 25

④ What is the mode score?

F 13
G 90.9
H 91
J 92

② Joanne bowled three games and averaged 134 points per game. If she bowled 125 for the first game and 145 for the second game, what was her score for the third game?

F 402
G 279
H 132
J 123

⑤ What is the median score?

A 13
B 90.9
C 91
D 92

⑥ What is the range for the ages of musicians in a community orchestra?

Ages of Musicians

20	52	42	59	46
48	64	36	21	40
14	45	38	29	34
39	35	61	19	27
26	18	52	57	

③ During May, the median house sale in the Carol Stream district was $134,095. There were 245 sales made in May. How many sales were below the median sale price?

A 122 sales
B 123 sales
C 245 sales
D 547 sales

F 64
G 50
H 38.5
J 14

Math Advantage Test Prep

65

Name _____

Practice Objective 43

DIRECTIONS
Read each question and choose the best answer. Then mark the space for the answer you have chosen.

SAMPLE
Kelly spins the spinner two times. What are the possible outcomes?

A 1-1, 1-2, 1-3, 2-1, 2-2, 2-3, 3-1, 3-2, 3-3
B 1-1, 1-2, 1-3, 2-2, 2-3, 3-3
C 2-2, 2-3, 3-3, 3-2, 3-1
D 1-1, 1-2, 1-3, 2-1, 2-2, 2-3, 4-1, 4-2, 4-3

① Denise spins this spinner once and tosses a coin once. What are the possible outcomes?

A 2 outcomes C 6 outcomes
B 4 outcomes D 8 outcomes

② Cari made a tree diagram to find the outcomes if she tosses a penny and a nickel. What possible outcome is missing?

F Heads (penny), Tails (nickel)
G Heads (penny), Heads (nickel)
H Tails (penny), Tails (nickel)
J Tails (penny), Heads (nickel)

③ Kendra tosses a number cube labeled from 1 to 6 and then tosses a coin. What are the possible outcomes?

A 12 outcomes C 6 outcomes
B 10 outcomes D 2 outcomes

④ Mike tosses two number cubes each labeled from 1 to 6. What are the possible outcomes?

F 6 outcomes   H 36 outcomes
G 12 outcomes J 72 outcomes

⑤ If you spin the spinner twice, which outcome is not possible?

A A sum of 10 C A sum of 18
B A sum of 14 D A sum of 4

⑥ If you spin the spinner twice, which outcome is not possible?

F A product of 3 H A product of 9
G A product of 6 J A product of 12

66 Math Advantage Test Prep

Name _____

Practice Objective 44

DIRECTIONS
Read each question and choose the best answer. Then mark the space for the answer you have chosen.

SAMPLE
Karl has 1 quarter, 6 dimes, 4 nickels, and 2 pennies in his pocket. If a coin falls out when he sits down, what is the probability that it will be a penny?

A $\frac{1}{2}$ C $\frac{2}{11}$
B $\frac{1}{3}$ D $\frac{2}{13}$

1 Jamie has 1 red marble, 3 white marbles, 4 blue marbles, and 2 green marbles in a pouch. If one is removed without looking, what is the probability that it will be red?

A 0.1   C 0.75
B 0.11   D 0.33

2 Kala has 3 red, 1 black, 2 green, and 2 blue T-shirts. She chose a shirt without looking. What is the probability that she chose a red T-shirt?

F $\frac{3}{4}$ H $\frac{1}{4}$
G $\frac{3}{8}$ J $\frac{1}{8}$

3 There are 4 answer choices for a problem-solving test question. What is the probability you can guess the correct answer?

A 0.1   C 0.2
B 0.17   D 0.25

4 Suppose Jesse tosses a number cube labeled 1, 3, 5, 7, 9, 11. What is P(5)?

F $\frac{1}{2}$ H $\frac{1}{4}$
G $\frac{1}{3}$ J $\frac{1}{6}$

5 Suppose Sandy tosses a number cube labeled from 3 to 8. What are Sandy's chances of rolling a 1 or a 2?

A 0   C $\frac{1}{2}$
B $\frac{1}{3}$ D 1

6 Use the spinner. What is P(2 or 4)?

F 0.25   H 0.60
G 0.50   J 0.75

7 What is the probability of hitting the shaded section of the dart target?

A $\frac{1}{4}$ C $\frac{1}{8}$
B $\frac{1}{3}$ D $\frac{1}{9}$

Math Advantage Test Prep   67

Name _____

Practice Objective 45

DIRECTIONS
Read each question and choose the best answer. Then mark the space for the answer you have chosen.

SAMPLE

David has a pair of black pants and a pair of brown pants. He also has a white shirt, a tan shirt, and a red shirt. How many different outfits can he make using one pair of pants and one shirt?

A 2 outfits
B 3 outfits
C 5 outfits
D 6 outfits

1 Maxine has a black skirt, a brown skirt, and a navy skirt. She also has a white blouse, a red blouse, and a yellow blouse. How many different outfits can she make using one skirt and one blouse?

A 6 outfits
B 8 outfits
C 9 outfits
D 12 outfits

2 Mona buys rye, white, wheat and pumpernickel bread. She also buys ham and roast beef. How many different kinds of sandwiches can she make using one type of bread and one type of meat?

F 4 kinds
G 6 kinds
H 8 kinds
J 10 kinds

3 Jim, Alex, Tom, Jack, and Bob play on a volleyball team. Two of these players were chosen to be co-captains. How many different pairs of co-captains are possible?

A 5 pairs
B 8 pairs
C 10 pairs
D 15 pairs

4 Carla, Dona, and Ruth volunteered to help Mr. Marshall. In how many ways can Mr. Marshall choose one person to record attendance and one person to take lunch money?

F 2 ways
G 3 ways
H 5 ways
J 6 ways

5 Pat has four new CDs. How many possible ways can he play the CDs if he always plays the same CD first?

A 3 possible ways
B 6 possible ways
C 12 possible ways
D 18 possible ways

Math Advantage Test Prep

Name _____

DIRECTIONS

Read each question and choose the best answer. Then mark the space for the answer you have chosen.

SAMPLE

Stuart's birthday is the day before Genevieve's birthday. Bobby's birthday is 4 days after Stuart's. Bobby's birthday is Dec. 31. When is Genevieve's birthday?

A Dec. 27
B Dec. 28
C Jan. 3
D Jan. 4

1 John's birthday is 3 days before Lizzie's. Lizzie's birthday is 5 days after Lorie's birthday. Lorie's birthday is Feb. 21. When is John's birthday?

A Feb. 21
B Feb. 23
C Feb. 24
D Feb. 26

2 Emma served pizza at her birthday party. Katy and Lisa each ate twice as much as Emma. Mary ate as much as Vicky and Lisa together. Patty ate as much as Mary. Vicky ate half as much as Lisa. Emma ate $\frac{1}{6}$ of a pizza. If they ate all the pizzas, what is the least number of pizzas Emma started with?

F 1 pizza
G 2 pizzas
H 3 pizzas
J 4 pizzas

For questions 3–6, use the map. Each side of the square equals one city block.

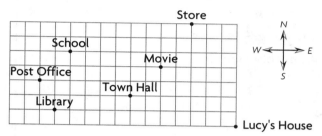

3 Lucy left her house and walked 7 blocks north and 3 blocks west. Where was she?

A Store
B School
C Movie
D Library

4 Ariel left the library and went 3 blocks north and 7 blocks east. Where was she?

F School
G Store
H Town Hall
J Movie

5 Scott left school and walked 2 blocks south, 4 blocks east, and 1 block south. Where was he?

A Movie
B Library
C Town Hall
D Store

6 Lucy left home and biked 10 blocks west, 5 blocks north, and 1 block west. Where was she?

F Store
G School
H Movie
J Library

Math Advantage Test Prep

Name _____

Practice Objective 46

DIRECTIONS
Read each question and choose the best answer. Then mark the space for the answer you have chosen.

SAMPLE

The sum of two numbers is 12. One number is $\frac{1}{2}$ of the other. What are the numbers?

A 4 and 8
B 7 and 5
C 3 and 6
D 9 and 3

1 Which two numbers have a sum of 52, a difference of 10, and a product of 651?

A 32 and 19
B 93 and 7
C 21 and 31
D 43 and 33

2 The sum of two numbers is 0.62, and the difference is 0.24. What are the two numbers?

F 0.36 and 0.12
G 0.40 and 0.22
H 0.43 and 0.19
J 0.44 and 0.18

3 Describe the rule that best describes the following pattern: 0.7, 1.4, 2.8, 5.6, …

A Add 0.7.
B Add 1.4.
C Multiply by 2.
D Multiply by 0.2.

4 Describe the rule that best describes the following pattern: $\frac{3}{8}, \frac{3}{4}, 1\frac{1}{8}, 1\frac{1}{2}, …$

F Divide by 2.
G Multiply by 2.
H Add $\frac{3}{16}$.
J Add $\frac{3}{8}$.

5 To begin her exercise training, Kara walked for 20 minutes on the first day, 25 minutes on the second day, and 30 minutes on the third day. If she continues at the same rate, on which day will she walk for 60 minutes?

A Day 6
B Day 7
C Day 8
D Day 9

6 Andy began jogging for 12 minutes for a week. The next week he increased his jogging time to 16 minutes. The third week he increased his jogging time to 20 minutes. At this rate, during which week can he expect to jog more 30 minutes?

F The 5th week
G The 6th week
H The 7th week
J The 10th week

70

Math Advantage Test Prep

Name _____

Practice Objective 46

DIRECTIONS
Read each question and choose the best answer. Then mark the space for the answer you have chosen.

SAMPLE
The Randolph family bought $4\frac{1}{8}$ pounds of green apples and $2\frac{9}{10}$ pounds of red apples. About how many pounds of apples did the family buy altogether?

A About 6 lb
B About 7 lb
C About 9 lb
D About 10 lb

1 On a camping trip, the scout troop hiked $13\frac{5}{8}$ mi on the first day. They hiked $9\frac{1}{10}$ mi on the second day. About how much farther did they hike the first day than the second day?

A About $3\frac{1}{2}$ mi
B About $4\frac{1}{2}$ mi
C About $5\frac{1}{2}$ mi
D About 21 mi

2 Each bus seats no more than 65 passengers. Will 6 buses be able to seat 360 students?

F Yes, because $360 \div 6 < 65$.
G Yes, because $6 \times 65 = 360$.
H No, because $65 < 360 \div 6$.
J No, because $6 \times 65 > 360$.

3 Allen takes trumpet lessons. Lessons were available on Monday, Tuesday, Thursday, or Friday, at 4:00 P.M. or at 5:00 P.M. How many choices did he have?

A 4 choices   C 8 choices
B 6 choices   D 10 choices

4 Notebooks are available in green, red, blue, black, or yellow. They come with or without an inside pocket for papers. How many choices are there?

F 7 choices   H 12 choices
G 10 choices   J 15 choices

5 How many different kinds of sandwiches can be made using wheat, white, or rye bread, either turkey or ham, and with or without cheese?

A 6 kinds of sandwiches
B 8 kinds of sandwiches
C 10 kinds of sandwiches
D 12 kinds of sandwiches

6 How many different ways can you have $0.75 using only quarters, dimes, and/or nickels?

F 18 ways   H 12 ways
G 14 ways   J 10 ways

Math Advantage Test Prep

Grade 6 • Harcourt Brace School Publishers

71

Name _____

Practice Objective 46

DIRECTIONS
Read each question and choose the best answer. Then mark the space for the answer you have chosen.

SAMPLE
Mrs. Land ordered 108 tickets for the concert, including 95 student tickets. How many adult tickets did she order?

A 213 tickets
B 203 tickets
C 23 tickets
D 13 tickets

1 The Lawrence family is planning a trip. If they travel at a rate of 60 miles per hour, how long will it take them to drive 270 miles?

A $3\frac{1}{2}$ hours
B 4 hours
C $4\frac{1}{2}$ hours
D $4\frac{3}{4}$ hours

2 Cindy is planning a trip. She plans to drive 180 miles the first day and $180 + y$ miles the second day. If $y = 60$, how far does she plan to drive the second day?

F 60 miles
G 120 miles
H 240 miles
J 360 miles

3 Sara is older than Sam by 4 years. Sara is 12 years old. If s = Sam's age, which equation would you use to find Sam's age?

A $s + 4 = 12$
B $s - 4 = 12$
C $s = 12$
D $4 \times s = 12$

4 After the first hour of the bake sale, 37 cupcakes were sold and 28 cupcakes were left. If t = the total number of cupcakes made for the sale, which equation would you use to find the total number made?

F $t = 37 - 28$
G $t + 28 = 37$
H $t - 37 = 28$
J $t = 28$

5 What is 90°F converted to degrees Celsius? Use the formula $C = \frac{5}{9} \times (F - 32)$. Round to the nearest degree.

A 12°C
B 50°C
C 32°C
D 64°C

Math Advantage Test Prep

Name _____

Practice Objective 47

DIRECTIONS
Read each question and choose the best answer. Then mark the space for the answer you have chosen. If a correct answer is *not here*, mark the space for NH.

SAMPLE
Use mental math to find
$80 + 159 + 20$.

A 200   C 260   E NH
B 259   D 300

1 Use mental math to find
$70 + 150 + 130$.

A 250   C 350   E NH
B 305   D 450

2 Use mental math to find $63 - 38$.

F 25   H 81   K NH
G 35   J 91

3 Use mental math to find 64×4.

A 266   C 246   E NH
B 256   D 224

4 Use mental math to find
$8 \times 10 \times 70$.

F 560   H 5,600   K NH
G 5,400   J 54,000

5 Use paper and pencil to find the remainder. $165 \div 7$

A 4   C 21   E NH
B 5   D 1,155

6 Use paper and pencil to find the product. 450×28

F 1,260   H 12,600   K NH
G 126   J 126,000

7 Use a calculator to find $\frac{5}{16}$ written as a decimal.

A 31.25   C 0.8   E NH
B 0.03125   D 0.3125

8 Use a calculator to find $3,901 \div 83$.

F 47   H 407   K NH
G 470   J 4

9 Anfernee earns $104.96 each day, including Saturdays and Sundays. At this rate, how much will he earn in a year? Which method is the most appropriate to answer the question?

A Mental math
B Paper and pencil
C Calculator
D Computer

Math Advantage Test Prep

Name _____

Practice Objective 48

DIRECTIONS
Read each question and choose the best answer. Then mark the space for the answer you have chosen.

SAMPLE

Marty paid $6.38 for bread and milk at the food store. How much change did he receive from $10.00?

A $16.38
B $4.62
C $4.38
D $3.62

1 While shopping for a birthday gift for his grandmother, Ken paid $12.50 for a plant and $1.75 for a card. How much change did he receive from a twenty-dollar bill?

A $5.75
B $6.75
C $7.75
D $14.25

2 Elizabeth bought a sweater for $18.95 and a skirt for $24.95. Sales tax was $3.51. What was the total amount that she paid?

F $57.41
G $47.41
H $43.90
J $40.39

3 Mr. Clakin's dry cleaning bill is $43.75. He gave the clerk three twenty-dollar bills. How much change should he receive?

A $103.75
B $60.00
C $16.25
D $6.25

4 The regular price of a pair of in-line skates is $95.00. They were placed on sale for 40% off. How much is the discount?

F $57.00
G $38.00
H $47.50
J $19.00

5 The regular price of a VCR is $212.60. Jana received a discount of 25%. How much did she pay for the VCR?

A $53.15
B $101.30
C $159.45
D $175.00

6 A bicycle costs $129. The sales tax is 6%. How much is the sales tax?

F $7.74
G $77.40
H $136.74
J $206.40

7 Art supplies are priced at $35. Sales tax is 7%. What is the total cost of these supplies?

A $2.45
B $32.55
C $37.45
D $45.00

74

Math Advantage Test Prep

Name _____

DIRECTIONS
Read each question and choose the best answer. Then mark the space for the answer you have chosen.

SAMPLE
A case of 24 cans of fruit juice costs $6.00. What is the cost per can of juice?

A $0.50
B $0.25
C $1.44
D $0.24

1 A 14-ounce box of breakfast cereal costs $2.66. What is the cost per ounce of the cereal?

A $0.19
B $0.29
C $0.37
D $0.27

2 Vicky bought 4 CDs at $12.99 each and 2 CDs at $14.99 each. How much did she have to pay altogether?

F $51.96
G $29.98
H $21.98
J $81.94

3 The Harris family bought a computer for $1,590 plus interest. They will make twelve payments of $145.75 each. How much was the interest?

A $159.00
B $145.75
C $1,749
D $132.50

4 Mrs. Scanlon paid $84.80 for 32 copies of a paperback novel for her reading class. What was the price for 10 books?

F $2,7136.00
G $36.50
H $26.50
J $2.65

5 For the picnic, Jillian bought cases of chips. Each case cost $6.72 and contains 24 individual bags. What was the cost for 12 bags?

A $0.28
B $0.38
C $3.36
D $4.56

6 What is the cost of borrowing $1,500 at 9% interest rate for 1 year?

F $11.25
G $13.50
H $135.00
J $162.00

7 Ms. Martin borrowed $2,000 at 12% interest per year. If she pays the loan back in two years, how much money must she repay?

A $240
B $480
C $2,240
D $2,480

Math Advantage Test Prep

75

Name _____

DIRECTIONS
Read each question and choose the best answer. Then mark the space for the answer you have chosen.

SAMPLE

Anna takes two piano lessons each week. What do you need to know in order to find out how much she pays her teacher each week?

A How long the lessons are
B What time the lessons are
C How much each lesson costs
D How much she practices

1 Larry is saving part of his weekly allowance for new ice skates. The skates cost $55.00 including the sales tax. What do you need to know in order to find out when he will be able to buy the skates?

A How much the sales tax is
B How much his allowance is
C How many weeks he will save his allowance
D How much he will save each week

2 It costs $4.25 for each ticket to a movie. Caitlyn has $20. What do you need to know to find out whether she can treat her entire family to a movie?

F How many tickets are available
G How many family members there are
H When the next show begins
J How much money is needed for the concession stand

3 Susan sews baby bibs to sell at craft fairs. Each bib takes no more than 10 minutes to make. What do you need to know to find out how long it will take her to sew the bibs for a special order?

A How many she sold at the last craft fair
B How many she will sew for the order
C How much she charges for each bib
D How many special orders she will have

4 Tony is saving to buy a new bike lock. He has saved $5.00 each week for 4 weeks. What do you need to know in order to find out when he will have enough money to buy the lock?

F How much the lock costs
G How much he earns each week
H How much other locks cost
J How much more he could save each week

5 Max and Cameron are going bowling. They plan to bowl two games at $1.50 each, rent shoes, and have $2.00 for snacks. What do you need to know in order to find out how much the activity will cost?

A What they will buy for snacks
B How long it will take to bowl two games
C How much they will pay to rent shoes
D What time they will start bowling

76 **Math Advantage Test Prep**

Name _____

Practice Objective 50

DIRECTIONS
Read each question and choose the best answer. Then mark the space for the answer you have chosen.

SAMPLE

Margie mixed 3 parts of yellow paint with 2 parts of blue paint. What is the ratio of parts of yellow paint to parts of blue paint?

A 3 to 5   C 2 to 3
B 2 to 5   D 3 to 2

1 The can of juice concentrate is diluted with 3 cans of water. What is the ratio of cans of concentrate to cans of water?

A 1 to 3   C 1 to 4
B 3 to 1   D 4 to 1

2 Katie earned $12 for babysitting 4 hours. What is the ratio of earnings to hours worked?

F $\frac{12}{4}$ H $\frac{1}{4}$
G $\frac{4}{12}$ J $\frac{1}{3}$

3 Michael's baseball team won 9 out of 12 games played. What is the ratio of games won to games played?

A $\frac{9}{3}$ C $\frac{3}{12}$
B $\frac{9}{12}$ D $\frac{12}{9}$

4 Kim's volleyball team lost 2 games out of 15 played. What is the ratio of losses to wins?

F 2 to 13   H 15 to 2
G 2 to 15   J 13 to 2

5 The U.S. flag has 13 stripes and 50 stars. What is the ratio of stars to stripes?

A 50 to 63   C 50 to 13
B 13 to 50   D 37 to 50

6 The sixth-grade class has 50 girls and 55 boys. What is the ratio of girls to boys?

F 50:105   H 55:50
G 105:50   J 50:55

7 Seven members of the Glee Club are girls. There are 17 members in all. What is the ratio of girls to all the members?

A 7:10   C 10:7
B 7:17   D 8:17

8 There are 15 members on Hubbard's softball team. The ratio of girls to boys is 1:2. How many boys are on the team?

F 5 boys   H 8 boys
G 7 boys   J 10 boys

Math Advantage Test Prep

Name _____

Practice Objective 50

DIRECTIONS
Read each question and choose the best answer. Then mark the space for the answer you have chosen.

SAMPLE

Alice walks 10 miles per week as part of her exercise program. Which proportion shows how many miles she walks in a year? (1 year = 52 weeks)

A $\dfrac{10}{1} = \dfrac{n}{52}$ C $\dfrac{10}{7} = \dfrac{52}{n}$

B $\dfrac{10}{1} = \dfrac{52}{n}$ D $\dfrac{7}{10} = \dfrac{52}{n}$

---

1 Russell sleeps 8 hours each day. Which proportion shows how many hours he sleeps in a week? (1 week = 7 days)

A $\dfrac{1}{8} = \dfrac{n}{7}$ C $\dfrac{8}{7} = \dfrac{1}{n}$

B $\dfrac{8}{1} = \dfrac{n}{7}$ D $\dfrac{7}{8} = \dfrac{1}{n}$

2 If Russell sleeps 8 hours each day, how many hours does he sleep in a week?

F 8 hours   H 56 hours
G 54 hours   J 48 hours

3 An 8-lb bag of cat food costs $6.40. Which proportion shows the cost per pound?

A $\dfrac{8}{6.40} = \dfrac{n}{1}$ C $\dfrac{6.40}{8} = \dfrac{n}{1}$

B $\dfrac{8}{n} = \dfrac{1}{6.40}$ D $\dfrac{6.40}{8} = \dfrac{1}{n}$

4 If an 8-lb bag of cat food costs $6.40, what is the cost per pound?

F $8   H $0.80
G $0.08   J $51.20

5 After exercising, Megan's pulse was 30 beats in 15 seconds. Which proportion shows the number of beats in 1 minute? (1 minute = 60 seconds)

A $\dfrac{30}{15} = \dfrac{n}{60}$ C $\dfrac{30}{60} = \dfrac{15}{n}$

B $\dfrac{15}{30} = \dfrac{n}{60}$ D $\dfrac{60}{15} = \dfrac{30}{n}$

6 If Megan's pulse was 30 beats in 15 seconds, how many times did it beat in 1 minute?

F 60 beats   H 120 beats
G 90 beats   J 150 beats

7 A map uses a scale of 3 cm for every 25 miles. If the map shows a distance of 15 cm, what is the actual distance?

A 5 miles   C 75 miles
B $8\dfrac{1}{4}$ miles   D 125 miles

Math Advantage Test Prep

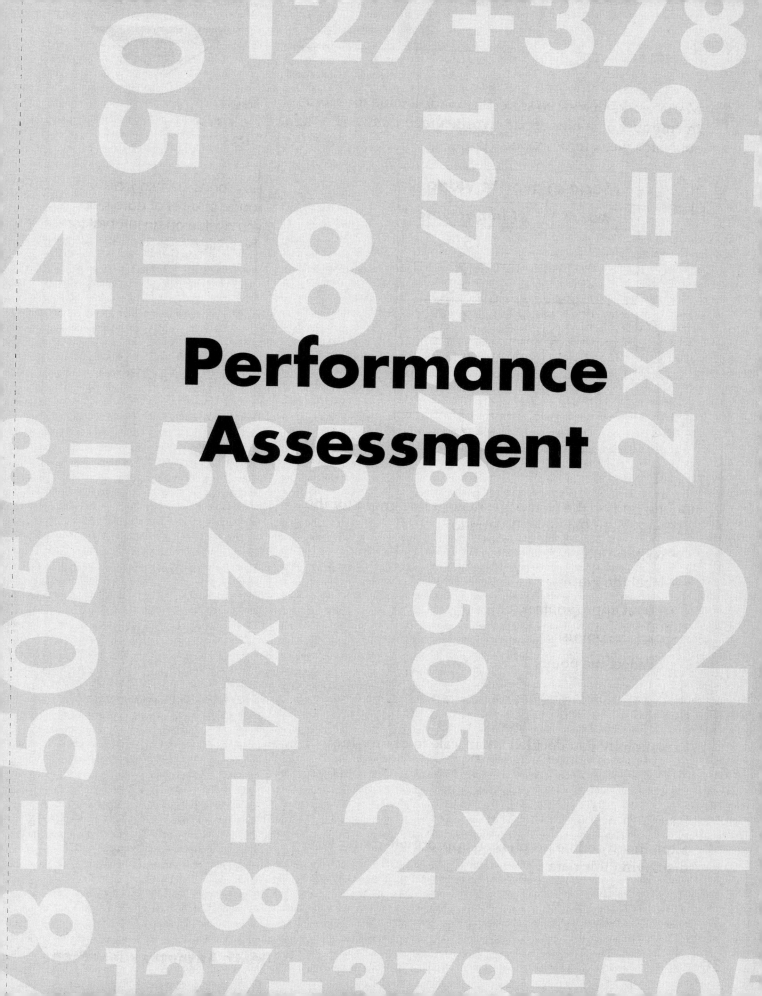

Performance Assessment

Name _____

1 Maria works in a video store. She made a table to record the number of copies of her favorite video that were sold each week for 6 weeks.

Maria's Favorite Videos	
Week	Number Sold
1	15
2	7
3	8
4	4
5	6
6	9

Part A

On the grid on the next page, make a line graph for the data shown in the table. Be sure to

- write a title
- label the graph
- use an appropriate scale
- plot the points
- connect the points

Part B

Explain how you decided what scale to use for the graph.

Part C

Explain why a line graph is an appropriate graph for displaying this data.

Test Taking Tips

How does knowing the range of a set of data help you decide on an interval for the scale?

Math Advantage Test Prep

80

Name _____

Practice Task 1

① Part A

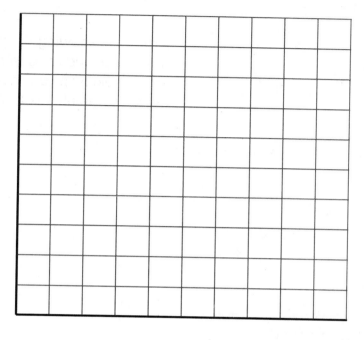

Part B

Explain your scale.

_____
_____
_____

Problem C

Explain why a line graph is appropriate.

_____
_____
_____
_____

Test Taking Tips

How can you check that your answers are accurate?

How can you check that your explanation is clear and complete?

Math Advantage Test Prep

2 Thomas works in a bakery.

He started this table to show the number of pound of apples he needs to bake medium apple pies.

Medium Apple Pie

Pounds of Apples	3				
Number of Pies	1	2	3	4	5

Part A

On the next page, complete the table for Thomas.

Part B

Thomas needs 4 pounds of apples to make a large apple pie. Make a table to show the number of pounds of apples he needs to make 1, 2, 3, 4, or 5 large apple pies.

Part C

Explain how you completed the tables. How are the two tables the same? How are they different?

Test Taking Tips

How does knowing Thomas needs 3 pounds of apples to make 1 pie help you know how to complete the table?

Math Advantage Test Prep

Name _____

② Part A

Complete the table.

Medium Apple Pie

Pounds of Apples	3				
Number of Pies	1	2	3	4	5

Part B

Make a table to show how many apples Thomas needs to make a large apple pie.

Part C

Use the data in the table to explain your answers.

_____

_____

_____

_____

_____

Practice Task 2

Test Taking Tips

How can you check that your answers are accurate?

How can you check that your explanation is clear and complete?

Name _____

③ Anthony is following directions to make a kite. The directions show a square with vertices labeled A, B, C, and D. Use the diagram to help you make a model of part of the kite.

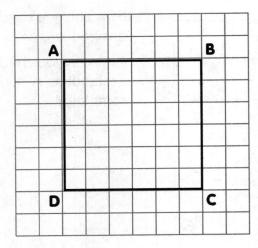

Test Taking Tips

What is made when you draw a line segment through a square?

Part A

On the grid on page 15, draw a large square. Label the vertices to match the diagram. Draw line segment AC. Describe the new polygons you have formed.

Part B

Now draw a large rectangle. Label the vertices E, F, G, H. Draw line segment EG. Describe the polygons you have formed.

Part C

Explain how the polygons you drew in the square and the polygons you drew in the rectangles are alike. Explain how they are different.

84

Math Advantage Test Prep

Name _____

Practice Task 3

3 Part A

Draw a large square and line segment in the grid.

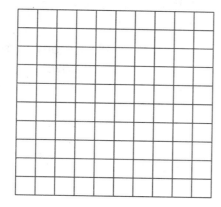

Describe the new polygons you have formed.

_____

Part B

Draw a rectangle and line segment in the grid.

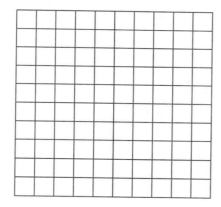

Describe the polygons you have formed.

_____

Part C

Explain the similarities and differences.

_____

_____

_____

Test Taking Tips

Math Advantage Test Prep 85

Name _____

Practice Task 4

④ Alex wants to cover some boxes with contact paper. One box is a cube with edges of 9 inches. He needs to know the sum of the area of the faces in order to find the total amount of paper he needs. Find the amount of contact paper Alex needs for this box.

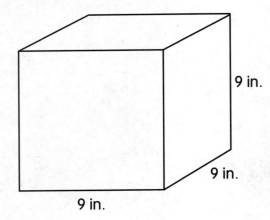

Test Taking Tips

How many faces does the box have?

Part A

Find the area of one face of the cube. Find the sum of the areas of faces. How much contact paper does Alex need?

Part B

Alex has another box that is a cube with sides of 10 inches. How much contact paper will Alex need to cover this box?

Part C

Explain how you found the sum of the areas of the faces on the cubes.

86

Math Advantage Test Prep

Name _____

Practice Task 4

④ Part A

How much contact paper does Alex need to cover a box that is a cube with 9-inch edges?

_____

Part B

How much contact paper does Alex need to cover a box that is a cube with 10-inch edges?

_____

Part C

Explain in words.

_____
_____
_____
_____
_____
_____
_____
_____

Test Taking Tips

How can you check that your answers are accurate?

How can you check that your explanation is clear and complete?

Math Advantage Test Prep   87

Name _____

Practice Task 5

5 At the circus, the people in every sixth row got free balloons.

The people in every fifteenth row got free ice cream.

Use the hundreds chart on the next page to help you find the rows where people got free balloons and free ice cream.

Part A

Circle the numbers in the hundreds chart for the rows where people got free balloons.

Mark X's on the numbers for the rows where people got free ice cream.

Part B

List the rows where people got free balloons.

List the rows where people got free ice cream.

In which rows did people get both free balloons and free ice cream?

Part C

Explain in words how you solved the problem.

Test Taking Tips

How do you decide which rows get free balloons?

88 Math Advantage Test Prep

Name _____

Practice Task 5

5 Part A

Circle the numbers in the hundreds chart to show the rows where people got free balloons. Mark X's on the chart to show the rows where people got free ice cream.

1	2	3	4	5	6	7	8	9	10
11	12	13	14	15	16	17	18	19	20
21	22	23	24	25	26	27	28	29	30
31	32	33	34	35	36	37	38	39	40
41	42	43	44	45	46	47	48	49	50
51	52	53	54	55	56	57	58	59	60
61	62	63	64	65	66	67	68	69	70
71	72	73	74	75	76	77	78	79	80
81	82	83	84	85	86	87	88	89	90
91	92	93	94	95	96	97	98	99	100

Part B

List the rows where people got free balloons.

_____

List the rows where people got free ice cream.

_____

List the rows where people got both free balloons and free ice cream.

_____

Part C

Explain in words.

_____

_____

_____

Test Taking Tips

Math Advantage Test Prep

6 The table shows Thomas's social studies test scores for his last seven tests.

Test	1	2	3	4	5	6	7
Score	67	49	75	50	85	80	49

His teacher said he had a choice of using the range, mean, median, or mode of his scores as his average grade for the tests.

This is the work you will do on the next page.

Part A

Find the range mean, median, and mode for the set of data. Which number should he use as his average grade?

Part B

Explain in words how you found the range, mean, median, and mode. Justify the number you chose as Thomas's average grade.

Practice Task 6

Test Taking Tips

How do you find the range, mean, median, and mode for a set of data?

Name _____

6 Part A

Find the range, mean, median, and mode for the set of test scores. Which number should he use?

Part B

Explain in words.

_____
_____
_____
_____
_____
_____
_____

Test Taking Tips

How can you check that your answers are accurate?

How can you check that your explanation is clear and complete?

7 Jose built a storage box that is 2 feet long, 2 feet wide, and 2 feet high. He needs to build a box that has a volume four times greater than the volume of this box.

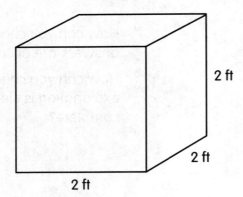

This is the work you will do on the next page.

Part A

Find the volume of the box with 2-foot edges.

Part B

Draw and label boxes with dimensions that have four times the volume of the smaller box.

Part C

Explain in words how changing the length, width, and height of the box changes its volume.

Practice Task 7

Test Taking Tips

How do you find the volume of a prism?

Name _____

7 Part A

Find the volume of the box with 2-foot edges.

Part B

Draw boxes which have a volume four times the volume of the smaller cube. Do not forget to label the dimensions.

Part C

Explain in words how changing the length, width, and height of the box changes its volume.

_____

_____

_____

Test Taking Tips

How can you check that your answers are accurate?

How can you check that your explanation is clear and complete?

Math Advantage Test Prep

Name _____

8 Carlos wants to find lines of symmetry in this figure.

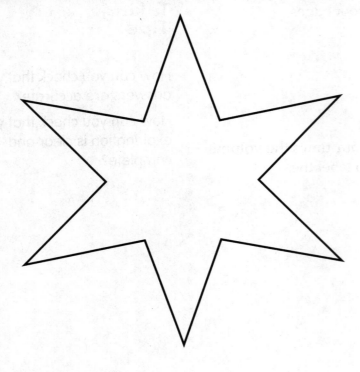

Test Taking Tips

How can you fold the figure so that the two halves are congruent?

Part A

Trace the figure. Fold it in half in different ways so that the two halves are congruent. Draw a line on the figure for each fold line that represents a line of symmetry.

Part B

Draw another figure that has one or more lines of symmetry. Draw the lines of symmetry for the figure.

Part C

Explain how you know a line drawn on a figure is a line of symmetry.

Math Advantage Test Prep

Name _____

8 Part A

Draw the lines of symmetry you found on this small star.
How many lines did you find?

_____

Part B

Draw another figure and show the lines of symmetry.

Part C

Explain how to determine the lines of symmetry in a figure.

_____
_____
_____
_____

Test Taking Tips

How can you check that your answers are accurate?

How can you check that your explanation is clear and complete?

Math Advantage Test Prep

Name _____

9. The sixth graders are having a pizza sale to make money for their outdoor-lab school. They have decided to make square pizzas. A pizza with 12-inch sides sells for $8.00. They need to decide how much to charge for 6-inch and 18-inch pizzas.

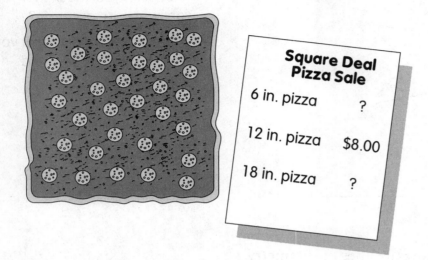

Square Deal Pizza Sale

6 in. pizza	?
12 in. pizza	$8.00
18 in. pizza	?

Test Taking Tips

How does finding the area of each pizza help you decide what to charge for the 6-inch and the 18-inch pizza?

Part A

On the next page, draw diagrams of the three square pizzas. Label the lengths of sides of each pizza. Find the area of each pizza. Explain how the areas are related.

Part B

Use the relationship between the areas of the pizzas to determine a price for the 6-inch and th 18-inch pizza.

Part C

Explain how you decided to charge for the pizzas.

96   **Math Advantage Test Prep**

Name _____

9 Part A

Draw diagrams of the three square pizzas. Be sure to label the sides and find the area of each.

Part B

Use the price of the 12-inch pizza and the relationship between the areas of the pizzas to decide on a price for the 6-inch pizza and the 18-inch pizza.

Part C

Explain in words.

_____

_____

_____

Practice Task 9

Test Taking Tips

How can you check that your answers are accurate?

How can you check that your explanation is clear and complete?

Name _____

10 As a homework assignment, Wendy is writing "real-life" problems for these expressions.

$1{,}170 \div 45 \qquad \dfrac{7}{8} \times 2$

$4\dfrac{3}{4} - \dfrac{7}{8} \qquad 10.75 + 15.49$

For the first expression Wendy wrote this problem.

There are 1,170 marbles in one jar.

How many marbles are in 45 jars?

Part A

Did Wendy write a problem appropriate for the expression? Explain on the next page.

Part B

Help Wendy with her homework. Write a problem for each expression.

Test Taking Tips

When do you use each operation in problems?

Math Advantage Test Prep

Name _____

10 Part A

Explain in words.

_____
_____
_____
_____

Part B

Write a "real-life" problem for each expression.

$1{,}170 \div 45$

$\frac{7}{8} x^2$

$4\frac{3}{4} - \frac{7}{8}$

$10.75 + 15.49$

Practice Task 10

Test Taking Tips

How can you check that your answers are accurate?

How can you check that your explanation is clear and complete?

Math Advantage Test Prep

Name _____

11 Jaime's little sister is selling lemonade. She has cups in three sizes. The small cups hold 6 ounces, the medium cups hold 12 ounces, and the large cups hold 15 ounces. Jaime is helping her sister decide how much to charge for the cups of lemonade. They decide to charge $0.40 for a medium cup.

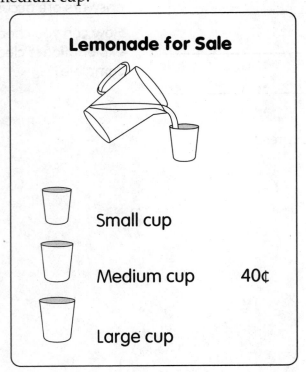

Test Taking Tips

How does knowing the capacity of each cup help you decide on a price?

Part A

On the next page, draw diagrams of the three cups. Write the number of ounces each cup holds on the cup. Explain how the capacities are related.

Part B

Use the relationship between the capacities of the cups to help the girls determine a price for the small cup and the large cup of lemonade.

Part C

Explain how you decided what to charge for each cup.

Name _____

11 Part A

Draw diagrams of the three cups. Be sure to label the cups.

Part B

Use the price of the medium cup of lemonade and the relationship between the capacities of the cups to decide on a price for the small cup and the large cup of lemonade.

Part C

Explain your reasoning.

_____

_____

_____

_____

Practice Task 11

Test Taking Tips

How can you check that your answers are accurate?

How can you check that your explanation is clear and complete?

Math Advantage Test Prep

101

12. Tessellations are designs that are found in nature, in art, and in manufactured objects. The design is a repeating arrangement of shapes that completely covers a plane with no gaps and no overlaps.

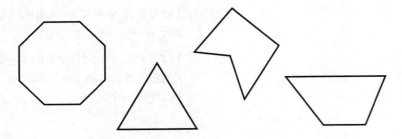

Part A

Trace each shape above several times and cut out your tracings. On the next page, circle the two shapes that can be used to form tessellations.

Part B

Draw a design that makes a tessellation. Use one of the shapes above or a different shape. Make at least three rows in your design.

Part C

Explain how you know your design forms a tessellation.

Test Taking Tips

What is a tessellation?

Name _____

12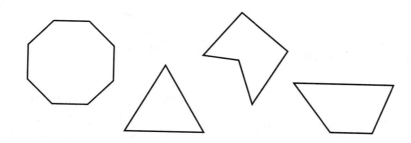

Test Taking Tips

How can you check that your answers are accurate?

How can you check that your explanation is clear and complete?

Part A

Which two shapes can be used to form tessellations?

_____

_____

Part B

Draw a design that makes a tessellation.

Part C

Explain in words.

_____

_____

_____

_____

Math Advantage Test Prep

Name _____

13 Beth has a 3-inch by 4-inch photo that she wants to enlarge and place in a frame that has a 12-inch by 12-inch opening. The original photo and the enlargement will be similar rectangles.

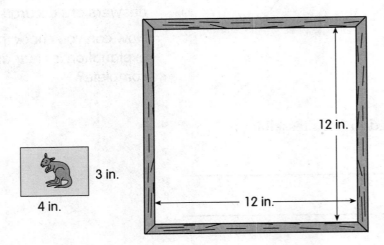

Here is the work you will do on the following page.

Part A

What are the dimensions of the largest enlargement Beth can fit in the frame? Draw a diagram to show the photo in the frame. Write the dimensions on the photo.

Part B

Suppose Beth wants to leave a space that is at least 2 inches on all sides between the photo and the frame. What are the dimensions of the largest enlargement Beth can use now? Draw a diagram to show the photo in the frame. Write the dimensions on the photo.

Part C

Explain how you found the dimensions of the photos.

Test Taking Tips

What do you know about the ratios of the lengths of matching sides in similar figures?

Math Advantage Test Prep

13 Part A

What are the dimensions of the largest enlargement Beth can fit in the frame? Don't forget to draw and label a diagram of the photo and frame.

Part B

What are the dimensions of the largest enlargement Beth can fit in the frame if she leaves at least 2 inches on all sides between the photo and the frame? Don't forget to draw and label a diagram of the photo and frame.

Part C

Explain in words.

_____

_____

_____

_____

Test Taking Tips

How can you check that your answers are accurate?

How can you check that your explanation is clear and complete?

14 Edward and his dad are putting together a new bicycle for Edward's little brother. There is a scale drawing in the instructions. It says the scale is $\frac{1}{2}$ inch = 1 foot.

Scale: $\frac{1}{2}$ inch = 1 foot

A scale of $\frac{1}{2}$ inch = 1 foot means that every $\frac{1}{2}$ inch in drawing of the bicycle represents 1 foot of actual size.

On the next page, you will find the actual size of the bicycle they are building. The steps are listed below.

Part A

Use an inch ruler to measure the height and the length of the bike in the diagram.

Part B

Use the height and length of the bicycle in the diagram and the scale to find the actual height and length of the new bicycle.

Part C

Explain how you used the diagram and the scale to find the actual size of the bicycle. Explain your reasoning.

Test Taking Tips

If the scale is $\frac{1}{2}$ inch = 1 foot, how do you find the actual length of something that measures 1 inch in a diagram?

14 Part A

Measure the height and the length of the bicycle in the diagram.

height in drawing: _____

length in drawing: _____

Part B

Find the actual height and length of the bicycle.

actual height: _____

actual length: _____

Part C

_____
_____
_____
_____
_____
_____
_____

Test Taking Tips

How can you check that your answers are accurate?

How can you check that your explanation is clear and complete?

Math Advantage Test Prep

Name _____

15 Joseph would like to be a cartoon animator when he grows up. However, he just found out that one animated movie may have up to 345,000 separate drawings. A $3\frac{1}{2}$-second scene takes about 100 drawings.

Joseph started this table to show about how many drawings would be made for every $3\frac{1}{2}$ seconds of animation.

Length of Film (in seconds)	Number of Drawings
$3\frac{1}{2}$	100
7	200
$10\frac{1}{2}$	300
	400
	500
	600
	700
	800
	900
	1,000

Part A

Copy and complete the table. How long is an animated film with 1,000 drawings?

Part B

If it takes 2 hours to complete a drawing, about how many hours will it take to make the drawings for an animated film that is 1 minute and 45 seconds (105 seconds) long?

Part C

Explain how you determined your answers.

Test Taking Tips

How many drawing are drawn for $3\frac{1}{2}$ seconds? For 7 seconds?

108   **Math Advantage Test Prep**

15 Part A

Copy and complete the table. How long is an animated film with 1,000 drawings?

Part B

If it takes 2 hours to complete a drawing, about how many hours will it take to make the drawings for an animated film that is 1 minute and 45 seconds (105 seconds) long?

_____

Part C

Explain how you determined your answers.

_____

_____

_____

_____

Test Taking Tips

How can you check that your answers are accurate?

How can you check that your explanation is clear and complete?

Math Advantage Test Prep

Name _____

16 Joe has designed this spinner for a new math board game he and his classmates are making. Before he decides on the rules for moving along the path on the game board, he wants to know the probability of spinning certain numbers.

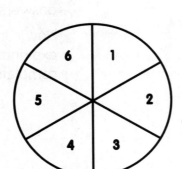

Part A

Use Joe's spinner. Write each probability as a fraction.

a. P (6)
b. P (3 or 4)
c. P (a multiple of 2)
d. P (odd number)
e. P (number less than 7)
f. P (number greater than 7)

Part B

Design a new 6-section spinner. Write numbers in the sections of your spinner so that all these probabilities are true.

a. P (even number) = $\frac{3}{6} = \frac{1}{2}$

b. P (number less than 4) = $\frac{0}{6} = 0$

c. P (7 or 8) = $\frac{2}{6} = \frac{1}{3}$

d. P (number greater than 12) = $\frac{0}{6} = 0$

Part C

Write three more probabilities that are true for the spinner you designed.

Test Taking Tips

Why is knowing probability important when you want to design a fair game?

Name _____

16 Part A

Write each probability as a fraction.

a. _____   d. _____

b. _____   e. _____

c. _____   f. _____

Test Taking Tips

How can you check that your answers are accurate?

How can you check that your explanation is clear and complete?

Part B

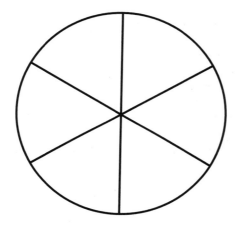

Design the new spinner. Be sure that all four probabilities are true for your spinner.

Part C

Write three more probabilities that are true for the spinner you designed.

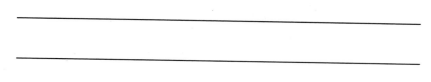

Math Advantage Test Prep 111

17 Jamaal bakes chocolate chip cookies and sells them to a grocery store. He sells each package of cookies to the store for $4.50. Each package costs Jamaal $2.50 to make and package. The difference between what he sells the cookies for and what they cost him to make and package is his profit. Here is a chart that he started.

Jamaal's Cookie Sales

Month	Packages Sold	Sales	Costs	Profits
January	3	$13.50	$7.50	$6
February	6			
March	5			
April	12			
May	8			
June	15			

Test Taking Tips

What do you need to figure out to solve the problem?

Part A

Complete the table showing Jamaal's cookies sales over a 6-month period.

Part B

Jamaal wants to make a profit of $36 each month. Write and solve an equation to show how many packages of cookies he will have to sell to make a profit of $36.
Let c = the number of packages of cookies he must sell.

Name _____

17 Part A

Complete the table showing Jamaal's cookie sales over a 6-month period.

Jamaal's Cookie Sales

Month	Packages Sold	Sales	Costs	Profits
January	3	$13.50	$7.50	$6
February	6			
March	5			
April	12			
May	8			
June	15			

Part B

Jamaal wants to make a profit of $36 each month. Write and solve an equation to show how many packages of cookies he will have to sell to make a profit of $36.
Let c = the number of packages of cookies he must sell.

Test Taking Tips

How can you check that your answers are accurate?

How can you check that your explanation is clear and complete?

18

Some Geometric Terms		
ray	line segment	intersecting
angle	parallel	acute
line	perpendicular	right
plane	point	obtuse

People often use geometric terms such as those above to describe figures on a map. Sara drew this map of her neighborhood.

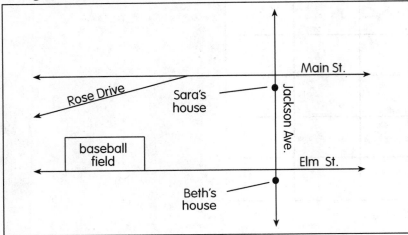

Part A

Use a geometric term to name the figure suggested by each location listed below. You may be able to use more than one term for some examples.

    a. Sara's house
    b. straight path from Beth's house north
    c. path from Sara's house to Beth's house
    d. Main Street intersected by Jackson Avenue
    e. Main Street and Elm Street
    f. baseball field

Part B

Draw a simple map of your neighborhood. Be sure to include figures that can be named as angles and lines or line segments.

Part C

Use at least 6 geometric terms to describe geometric figures on your map.

Test Taking Tips

What object in your classroom reminds you of each of these geometric terms?

    angle

    line segment

    ray

Math Advantage Test Prep

Name _____

Practice Task 18

18 Part A

List a geometric term suggested by each location.

a. _____

b. _____

c. _____

d. _____

e. _____

f. _____

Part B

Draw a simple map of your neighborhood. Be sure to include figures that can be named as polygons, angles, and line segments.

Part C

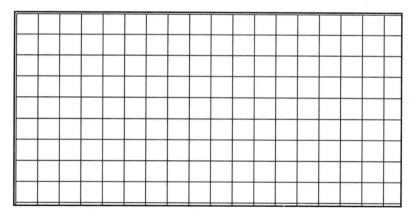

Identify six geometric figures on your map.

_____

_____

_____

_____

Test Taking Tips

How can you check that your answers are accurate?

How can you check that your explanation is clear and complete?

Math Advantage Test Prep   115

Name _____

19 Karen made this table to record her scores on math tests during the year.

Karen's Test Scores							
Test	1	2	3	4	5	6	7
Scores	88	90	74	94	100	80	90

She wants to analyze her scores to see how she is doing.

Part A

Make a graph for the data shown in the table. Remember to label the graph and write a title.

Part B

Find the mean, median, and mode of Karen's test scores.

Part C

Explain your choice for the type of graph you made to display Karen's scores. Write a sentence to describe the data.

Test Taking Tips

When is it most appropriate to use a bar graph? a circle graph? a line graph?

116 Math Advantage Test Prep

Name _____

19 Part A

Decide what kind of graph will best display the data. Make the graph on the grid. Be sure to label the graph and write a title.

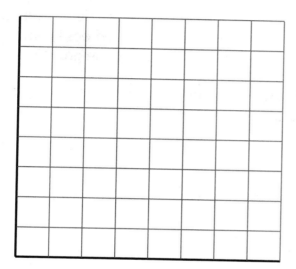

Part B

Find the mean, median, and mode of Karen's scores.

Mean: _____

Median: _____

Mode: _____

Part C

Explain your choice for the type of graph you made to display Karen's scores. Write a sentence to describe the data.

_____

_____

_____

Test Taking Tips

How can you check that your answers are accurate?

How can you check that your explanation is clear and complete?

Math Advantage Test Prep

Name _____

20 Juan conducted a survey of 20 students in his class about their favorite subjects in school. His survey revealed the following information:

    10 students preferred social studies
    5 students preferred art
    3 students preferred science
    1 student preferred English
    1 student preferred math

Part A

Draw a circle graph to illustrate Juan's survey results. Be sure to label the sections of your graph and write a title.

Part B

Draw a bar graph to illustrate Juan's survey results. Be sure to label your graph and write a title.

Part C

Which graph do you think is easier to understand? Explain your choice.

Test Taking Tips

What type of graph can be used to show how the parts relate to the whole? How do you determine the number of degrees in each section of a circle graph?

Math Advantage Test Prep

Name _____

20 Part A

Draw a circle graph. Remember labels and a title.

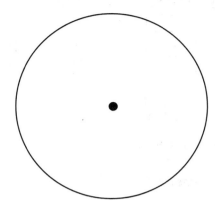

Part B

Draw a bar graph. Remember labels and a title.

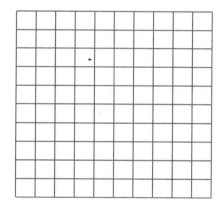

Part C

Which graph do you think is easier to understand?
Explain your choice.

_____

_____

Test Taking Tips

How can you check that your answers are accurate?

How can you check that your explanation is clear and complete?

㉑

Math Test-Score Data				
Tests	Range	Median	Mode	Mean
37, 65, 42, 55, 65	28	55	65	53
58, 42, 75, 85, 45	43	58	—	61
38, 27, 30, 55, 45	28			
45, 35, 20, 25, 45		35		

Test Taking Tips

How do you find range, median, mode and mean?

The table above shows the scores four students received on the last five math tests in Mr. Wong's class. All four of these students usually have an average of between 95 and 100 in math.

Part A

Analyze the test-score data by completing the table of data.

Part B

Compare the scores the students received on these tests with how they usually score on math tests. What do you think this comparison shows? Justify your answers on the next page.

Math Advantage Test Prep

Name _____

Practice Task 21

21 Part A

Complete the table of data.

Math Test-Score Data				
Tests	Range	Median	Mode	Mean
37, 65, 42, 55, 65	28	55	65	53
58, 42, 75, 85, 45	43	58	—	61
38, 27, 30, 55, 45	28			
45, 35, 20, 25, 45		35		

Part B

Compare the scores the students received on these tests with how they usually score on math tests. What do you think this comparison shows?

Test Taking Tips

How can you check that your answers are accurate?

How can you check that your explanation is clear and complete?

_____
_____
_____
_____

Math Advantage Test Prep 121

Name _____

22. Adam is planning to drive from his home in Cleveland, Ohio to Seattle, Washington, a distance of 2,310 miles. He estimates that his car averages 30 miles per gallon of gas and that gas costs about $1.25 per gallon. He plans to drive between 400 miles and 500 miles each day.

Part A

About how many gallons of gas will Adam use on his trip? Write an equation and solve.

About how much will the gas for the trip cost? Write an equation and solve.

Part B

How many days will it take Adam to drive

- if he travels 400 miles a day?
- if he travels 500 miles a day?

Write equations to solve.

Part C

If Adam drives 500 miles each day for the first two days and 450 miles each day for the next two days, how many miles will he have left to drive? Draw a diagram to solve.

Test Taking Tips

What information does the problem give you?

22 Part A

Write an equation that you could use to determine how many gallons of gas Adam will use on his trip. Solve the equation.

_____

_____

Write an equation that you could use to determine how much Adam will spend on gas during his trip. Solve your equation.

_____

_____

Part B

Write equations that you could use to determine how many days it will take Adam to make the trip

- if he drives 400 miles a day.
- if he drives 500 miles a day.

Solve the equations.

_____

_____

Part C

Draw a diagram you could use to help you determine how many miles Adam will have left to drive if he drives 500 miles each day for the first two days and 450 miles each day for the next two days.

Test Taking Tips

How can you check that your answers are accurate?

How can you check that your explanation is clear and complete?

Math Advantage Test Prep

Name _____

Practice Task 23

23 The Acme Storage Company is building a new warehouse. The floor plan below shows the length and width of the building. The building has 14 foot ceilings. The company needs a building with an area of at least 5,000 ft² and a volume of 60,000 ft³.

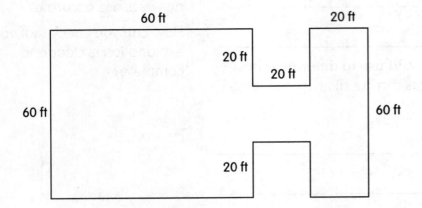

Part A

What is the area of the building? What is the volume of the building? Does the building meet the area and volume needs of the company? Show your work.

Part B

Draw a different floor plan that would meet the space needs of the Acme Storage Company. Label the dimensions on your plan.

Part C

Explain how you know that your floor plan will meet the space needs of the company.

Test Taking Tips

How can knowing the formula for the area of a rectangle help you find the area of the floor of the building?

Math Advantage Test Prep

Name _____

Practice Task 23

23 Part A

Find the area and the volume of the building shown on the diagram. Be sure to show your work.

area:

_____

volume:

_____

Part B

Draw another floor plan to meet the area needs of the company. Don't forget to label the length and width.

Part C

Explain how you know that your floor plan will meet the space needs of the company.

_____

_____

_____

_____

Test Taking Tips

How can you check that your answers are accurate?

How can you check that your explanation is clear and complete?

Math Advantage Test Prep

24 Jolene recorded these temperature highs and lows for the past four days.

Temperatures in Centerville					
	Mon	Tue	Wed	Thu	Fri
High	23°C	21°C	19°C	18°C	
Low	5°C	8°C	9°C	10°C	

Part A

Use the data in the table to make a multiple-line graph. Be sure to:

- choose an appropriate temperature scale.
- label the axes of the graph.
- write a title.
- include a key.
- mark a point for each high temperature and connect the points.
- mark a point for each low temperature and connect the points.

Part B

If the temperature trends continue, what do you think the temperature high and the temperature low will be on Friday?

Part C

Explain how you decided on your prediction.

Test Taking Tips

What type of graph do you use to show a change over time between two sets of data?

Math Advantage Test Prep

Name _____

24

Part A

Make a multiple-line graph of the temperature data in the table. Label your graph and write a title.

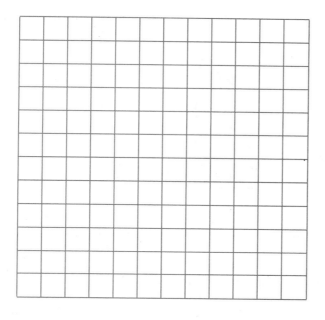

Part B

Predict the temperature high and low on Friday.

_____

Part C

Explain how you decided on your prediction.

_____

_____

_____

_____

_____

Test Taking Tips

How can you check that your answers are accurate?

How can you check that your explanation is clear and complete?

Math Advantage Test Prep

25 Jason and his friends like to play football. The line at the center of the field is called the 50-yard line. The lines at the ends of the field are called goal lines.

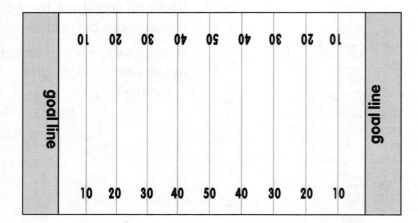

Test Taking Tips

How can you use integers to show the total number of yards gained or lost?

Part A

At one point in the game, Jason's team had the ball at the 50-yard line. They wanted to move toward the goal line at the right. In the next three plays they gained 5 yards, lost 3 yards, and gained 14 yards. Where was the ball after the three plays? Draw a picture or write an explanation to justify your answer.

Part B

Later in the game, Ted's team had the ball on the 50-yard line. They wanted to move toward the goal line at the left. After 3 plays they were 10 yards closer to the goal at the left. During the first play they had lost yards and during the next two plays they gained yards. Describe the number of yards they could have lost and gained in the three plays. Draw a picture and write an explanation to justify your answer.

Math Advantage Test Prep

Name _____

Practice Task 25

25 Part A

Draw a diagram or write an explanation to describe how the ball moved in 3 plays for Jason's team. Don't forget to show where the ball was at the end of the 3 plays.

Part B

Draw a diagram and write an explanation to describe how the ball could have moved in 3 plays for Ted's team.

Test Taking Tips

How can you check that your answers are accurate?

How can you check that your explanation is clear and complete?

Math Advantage Test Prep

129

26 The playoffs for the Little League division champs are between the Giants and the Marlins. The series is based on the best out of 5. That means that the team that wins the most games out of 5 possible games will be the division champs.

Complete the following on the answer sheet.

Part A

Make an organized list of all possible outcomes. Include wins, losses, and ties.

Part B

What is the least number of games that need to be played to determine the champs? What is the greatest number of games that can be played to determine the champs? Explain your answer.

Part C

Suppose the Marlins won the playoffs in 4 games. One way you can show the games they could have won is to list the winners of each of the 4 games like this:

Marlins, Marlins, Giants, Marlins.

This means the Marlins won games 1, 2, and 4, and the Giants won game 3. List other ways the Marlins could have won the playoffs in 4 games.

List the possible winners in a best out of 5 playoff where the Marlins win in 4 games.

Test Taking Tips

What will happen if the Marlins win the first 3 games?

26 Part A

Make an organized list of all possible outcomes. Include wins, losses, and ties.

Part B

What is the least number of games played in a best out of 5 playoff?

_____

What is the greatest number of games played?

_____

Explain your answer.

_____

_____

_____

Part C

List the possible winners in a best out of 5 playoff where the Marlins win in 4 games.

_____

_____

_____

Test Taking Tips

How can you check that your answers are accurate?

How can you check that your explanation is clear and complete?

27 Anita deposited $75.00 in a savings account. The account pays 3% simple interest. She plans to leave the money in the bank for 5 years.

Part A

Write an equation to show how much money Anita will have in the account after 1 year. (Remember, she will have the amount she originally deposited plus the interest she has earned on the amount she deposited.)

Solve the equation.

Part B

Complete this table to show the interest that will be earned on Anita's money after it has been in the savings account for 1 year through 5 years.

Anita's Interest					
Years in Account	1	2	3	4	5
Simple Interest Earned	$2.25				

Part C

Describe a pattern you see in the table.

Test Taking Tips

What information does the problem give you?

What do you need to find out to solve the problem?

How do you find simple interest?

27 Part A

Write and solve an equation to show how much money Anita will have in her savings account after 1 year.

_____

_____

_____

_____

Part B

Complete the table.

Anita's Interest					
Years in Account	1	2	3	4	5
Simple Interest Earned	$2.25				

Part C

Describe a pattern you see in the table.

_____

_____

_____

_____

_____

Test Taking Tips

How can you check that your answers are accurate?

How can you check that your explanation is clear and complete?

Math Advantage Test Prep

Name _____

28 Miguel is transforming figures on a coordinate plane to make interesting designs. He starts with triangle ABC with vertices at (2, 1), (5, 1), and (2, 3).

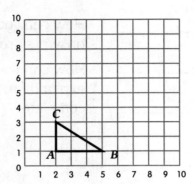

Part A

On the grid, draw the coordinate plane shown above. Draw triangle *ABC*. Then reflect triangle *ABC* across the *x*-axis. Draw the new figure. Label its vertices *A'*, *B'*, and *C'*. What are the coordinates of the vertices of the new triangle?

Part B

Compare the coordinates of triangle *ABC* and of triangle *A'B'C'*. Explain how the coordinates are alike and how they are different.

Test Taking Tips

How do you locate a point on a coordinate plane?

Math Advantage Test Prep

Name _____

28 Part A

On the coordinates below, draw triangle ABC. Then reflect triangle ABC over the x-axis and label the new figure triangle A'B'C'.

Test Taking Tips

How can you check that your answers are accurate?

How can you check that your explanation is clear and complete?

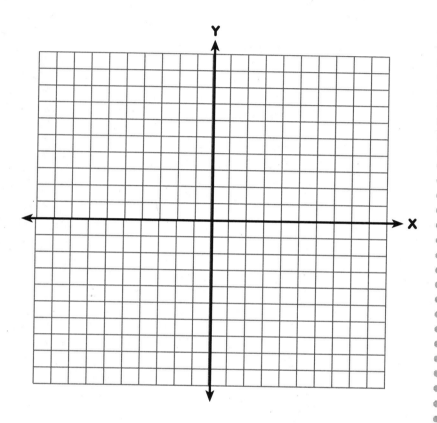

Part B

Compare the coordinates of triangle ABC and of triangle A'B'C'. Explain how the coordinates are alike and how they are different.

Math Advantage Test Prep